Advanced Self-Cleaning Surfaces

Advanced Self-Cleaning Surfaces

Editors

Maria Vittoria Diamanti
Massimiliano D'Arienzo
Carlo Antonini
Michele Ferrari

Basel • Beijing • Wuhan • Barcelona • Belgrade • Novi Sad • Cluj • Manchester

Editors

Maria Vittoria Diamanti
Politecnico di Milano
Milan
Italy

Massimiliano D'Arienzo
University of Milano-Bicocca
Milan
Italy

Carlo Antonini
University of Milano-Bicocca
Milan
Italy

Michele Ferrari
National Research Council,
CNR-ICMATE
Rome
Italy

Editorial Office
MDPI
St. Alban-Anlage 66
4052 Basel, Switzerland

This is a reprint of articles from the Topic published online in the open access journals *Coatings* (ISSN 2079-6412), *Materials* (ISSN 1996-1944), *Membranes* (ISSN 2077-0375), and *Nanomaterials* (ISSN 2079-4991) (available at: http://www.mdpi.com).

For citation purposes, cite each article independently as indicated on the article page online and as indicated below:

Lastname, A.A.; Lastname, B.B. Article Title. *Journal Name* **Year**, *Volume Number*, Page Range.

ISBN 978-3-7258-0231-9 (Hbk)
ISBN 978-3-7258-0232-6 (PDF)
doi.org/10.3390/books978-3-7258-0232-6

Cover image courtesy of Carlo Antonini

© 2024 by the authors. Articles in this book are Open Access and distributed under the Creative Commons Attribution (CC BY) license. The book as a whole is distributed by MDPI under the terms and conditions of the Creative Commons Attribution-NonCommercial-NoDerivs (CC BY-NC-ND) license.

Contents

About the Editors vii

Carlo Antonini, Massimiliano D'Arienzo, Michele Ferrari and Maria Vittoria Diamanti
Advanced Self-Cleaning Surfaces
Reprinted from: *Materials* **2024**, *17*, 537, doi:10.3390/ ma17030537 1

Jie Luo, Yi Xu, Hongqiang Chu, Lu Yang, Zijian Song, Weizhun Jin, et al.
Research on the Performance of Superhydrophobic Cement-Based Materials Based on Composite Hydrophobic Agents
Reprinted from: *Materials* **2023**, *16*, 6592, doi:10.3390/ ma16196592 4

Dongik Cho and Jun Kyun Oh
Silica Nanoparticle-Infused Omniphobic Polyurethane Foam with Bacterial Anti-Adhesion and Antifouling Properties for Hygiene Purposes
Reprinted from: *Nanomaterials* **2023**, *13*, 2035, doi:10.3390/nano13142035 21

Nuo Chen, Kexin Sun, Huicong Liang, Bingyan Xu, Si Wu, Qi Zhang, et al.
Novel Engineered Carbon Cloth-Based Self-Cleaning Membrane for High-Efficiency Oil–Water Separation
Reprinted from: *Nanomaterials* **2023**, *13*, 624, doi:10.3390/nano13040624 38

Hannelore Peeters, Silvia Lenaerts and Sammy W. Verbruggen
Benchmarking the Photocatalytic Self-Cleaning Activity of Industrial and Experimental Materials with *ISO 27448:2009*
Reprinted from: *Materials* **2023**, *16*, 1119, doi:10.3390/ma16031119 53

Chunling Zhang, Yichen Yang, Shuai Luo, Chunzu Cheng, Shuli Wang and Bo Liu
Fabrication of Superhydrophobic Composite Membranes with Honeycomb Porous Structure for Oil/Water Separation
Reprinted from: *Coatings* **2022**, *12*, 1698, doi:10.3390/coatings12111698 66

Raziyeh Akbari, Mohammad Reza Mohammadizadeh, Carlo Antonini, Frédéric Guittard and Thierry Darmanin
Controlling Morphology and Wettability of Intrinsically Superhydrophobic Copper-Based Surfaces by Electrodeposition
Reprinted from: *Coatings* **2022**, *12*, 1260, doi:10.3390/coatings12091260 81

Yuting Zhang, Tingping Lei, Shuangmin Li, Xiaomei Cai, Zhiyuan Hu, Weibin Wu, et al.
Candle Soot-Based Electrosprayed Superhydrophobic Coatings for Self-Cleaning, Anti-Corrosion and Oil/Water Separation
Reprinted from: *Materials* **2022**, *15*, 5300, doi:10.3390/ma15155300 97

Xiao Wang, Cheng Fu, Chunlai Zhang, Zhengyao Qiu and Bo Wang
A Comprehensive Review of Wetting Transition Mechanism on the Surfaces of Microstructures from Theory and Testing Methods
Reprinted from: *Materials* **2022**, *15*, 4747, doi:10.3390/ma15144747 106

Fabrice Noël Hakan Karabulut, Dhevesh Fomra, Günther Höfler, Naveen Ashok Chand and Gareth Wesley Beckermann
Virucidal and Bactericidal Filtration Media from Electrospun Polylactic Acid Nanofibres Capable of Protecting against COVID-19
Reprinted from: *Membranes* **2022**, *12*, 571, doi:10.3390/membranes12060571 127

Zhongyong Qiu and Chunju He
Polypropylene Hollow-Fiber Membrane Made Using the Dissolution-Induced Pores Method
Reprinted from: *Membranes* **2022**, *12*, 384, doi:10.3390/membranes12040384 . . . . **141**

About the Editors

Maria Vittoria Diamanti

Maria Vittoria Diamanti is an Associate Professor at the Department of Chemistry, Materials and Chemical Engineering, "Giulio Natta" of Politecnico di Milano, where she teaches science and technology of materials and the durability of cementitious and ceramic materials to BD and MD engineering courses. Her research interests are mostly related to the production, characterization and applications of nanostructured titanium oxides, either grown using electrochemical techniques on titanium and its alloys, or as nanoparticles or sol–gel applied to different substrates as coating or in bulk (polymers, glass, mortars, etc.). The versatility of titanium dioxide has allowed her during the years to achieve expertise in diverse fields—from self-cleaning construction materials to photocatalysts for wastewater treatment and air quality improvement, surface treatments for corrosion resistance, memristive materials for electronics and jewelry alloys—and to use a multidisciplinary approach in the production, characterization and applications of these oxides. She is also involved in research activities on the corrosion resistance of steel alloys in particular environments, such as chloride-containing environments, concrete and ionic liquids. She is an author of more than 70 articles published in international journals (h-index: 26, source: Scopus, Dec 2023), an editor of 3 books, an academic editor of Coatings and of Discover Chemical Engineering, and a guest editor of 5 Special Issues of international journals.

Massimiliano D'Arienzo

Massimiliano D'Arienzo is an Associate Professor of General and Inorganic Chemistry at the Department of Materials Science in the University of Milano-Bicocca since 2018. His expertise spans from the synthesis using soft chemistry methods of ceramics and inorganic materials with controlled morphological and surface features, to their exploitation in hybrid materials employed in a wide range of applications (e.g., photocatalysis, catalysis, LIBs, gas sensing and automotive). The results of his research activity provided scientific and technological impacts, leading to several financed national and European projects, more than 80 peer-reviewed papers, two book chapters and five patents.

Carlo Antonini

Carlo Antonini received a BSc in Aerospace Engineering (2004) and an MSc in Aeronautical Engineering (2007) from Politecnico di Milano, Italy, and a PhD in Technologies for Energy and Environment from University of Bergamo (2011), Italy, with a thesis titled "Superhydrophobicity as a strategy against icing". In 2012, he received support from the European Research Council (ERC) to join ETH Zurich, Switzerland, as a Marie Curie Fellow. In March 2015, he joined Swiss Federal Laboratory for Material Science and Technology (EMPA) as a scientist, focusing on the control of surface-wetting properties of cellulose-based materials for various engineering applications, ranging from liquid separation (oil remediation) to thermal insulation. In September 2018, he joined the Department of Materials Science at the University of Bicocca-Milano (Italy) with the support of the Rita Levi Montalcini Fellowship. He is currently an associate professor and leads the Surface Engineering and Fluid Interfaces Laboratory (SEFI Lab), with research interests in non-wetting surfaces, icephobicity, chitosan-based materials and coatings and 3D printing using digital light processing.

Michele Ferrari

Michele Ferrari is a Senior Researcher at CNR-ICMATE (Institute of Condensed Matter and Technologies for Energy) in Genova, Italy. In 1994, he joined the group of Dr. R. Miller at Max Planck Institute for Colloids and Interfaces (MPI_KG) in Berlin as an EU-HCM postdoc. He was a visiting researcher (2001) at MPI kg (Golm, Germany), MPI-PR (Mainz, Germany) (2006), UNISA (Adelaide, Australia) (2009) and INFLPR (Magurele, Romania) (2013). He is a member of IN2 Universitat de Barcelona, COST European Action networks in Physics of Droplets (P21), (CM1101) and (MP1106). His scientific interests include wetting, superhydrophobic and amphiphobic coatings with application in solar panels and marine environment, AFM in the studies of adsorption and self-assembly of amphiphiles at the liquid–solid interface and 3D confocal profilometry for the in vitro cell response screening of drug treatments. He is a Member of the ESA Topical Team on the interfacial properties of surfactant systems and microgravity experiments (onboard Shuttle and ISS). He is an author and co-author (Scopus h-index: 32) of more than 110 papers in peer-reviewed scientific journals, book chapters and encyclopedia voices. He has been an editor of many books and is currently a member of several editorial boards of journals and international conferences.

Editorial

Advanced Self-Cleaning Surfaces

Carlo Antonini [1], Massimiliano D'Arienzo [1], Michele Ferrari [2] and Maria Vittoria Diamanti [3,*]

[1] Department of Materials Science, University of Milano-Bicocca, 20125 Milan, Italy; carlo.antonini@unimib.it (C.A.); massimiliano.darienzo@unimib.it (M.D.)
[2] Institute of Condensed Matter Chemistry and Technologies for Energy, National Research Council, CNR-ICMATE, 16149 Genoa, Italy; michele.ferrari@ge.icmate.cnr.it
[3] Department of Chemistry, Materials and Chemical Engineering "Giulio Natta", Politecnico di Milano, 20131 Milan, Italy
* Correspondence: mariavittoria.diamanti@polimi.it

Hydrophobicity, olephobicity, hemophobicity, amphiphobicity, omniphobicity, icephobicity. In the last fifteen years, an exponentially increasing number of scientific studies have focused on surface micro- and nano-engineering to develop surfaces capable of repelling liquids (and even solids, in the case of ice) by controlling wetting properties. Such interest primarily derives from nature observations, as some plant leaves have developed peculiar wetting properties for self-cleaning purposes, e.g., to remain clean by preventing the adhesion of particles or bacteria on their surfaces [1]. The classic example is Nelumbo nucifera, more commonly known as lotus, but other plants such as Colocasia esculenta, Mutisia decurrens [2] and Salvinia molesta [3] possess similar properties. All these plants are characterized by superhydrophobicity, which combines high water repellence with high mobility: these properties are given by a combination of chemistry (leaves are coated with intrinsically hydrophobic wax) and surface hierarchical topography at the micro- and nano-scale.

Researchers first tried to replicate synthetic superhydrophobic surfaces, but then also extended the concept to all type of liquids (oil, low-surface tension fluids, blood) with an ambitious scientific question: can we design surfaces that can repel any type of liquid, and thus fabricate the perfect self-cleaning surface? The ambition is not a pure academic exercise, but it can have a significant impact on a plethora of applications, spanning from anti-icing to anti-fogging mirrors, low-drag ship hulls, anti-bacterial and anti-viral surfaces, and food containers.

However, the repelling of liquids is not the only strategy to achieve self-cleaning surfaces: the superhydrophilic principle has been adopted to produce self-cleaning glass for building envelopes and photovoltaics, and is already widely diffuse at an industrial level [4]. This effect originates from photoinduced phenomena, and relies on a combined surface hydroxylation and degradation of soiling agents [5].

It is thus clear how self-cleaning surfaces can be achieved by a variety of strategies, materials and technologies, and their applications are as wide as their production methods.

The current interdisciplinary topic, "Advanced Self-Cleaning Surfaces", has collected a variety of works dedicated to self-cleaning, from theoretical approaches to small-scale laboratory experiments and material validation in relevant environments, demonstrating how tailoring the surface at nano- and micro-meter scale surface phenomena can affect larger scale phenomena. We believe that joining all aspects of such a multifaceted phenomenon into a single collection of articles will help the scientific community involved in this field to improve collaboration among diverse disciplines and promote novel insights across interdisciplinary research fields. On account of its peculiar interdisciplinary character, this article collection was published across different journals: *Materials*, *Membranes*, *Nanomaterials* and *Coatings*. The ten articles included in the topical collection highlight

Citation: Antonini, C.; D'Arienzo, M.; Ferrari, M.; Diamanti, M.V. Advanced Self-Cleaning Surfaces. *Materials* **2024**, *17*, 537. https://doi.org/10.3390/ma17030537

Received: 5 January 2024
Accepted: 18 January 2024
Published: 23 January 2024

Copyright: © 2024 by the authors. Licensee MDPI, Basel, Switzerland. This article is an open access article distributed under the terms and conditions of the Creative Commons Attribution (CC BY) license (https://creativecommons.org/licenses/by/4.0/).

diverse material compositions, production methodologies and applications, as discussed in the following.

Oil–water separation is confirmed to be a hot topic, and has been addressed by several research works with different approaches. A PVDF/DP8/SiO$_2$ composite honeycomb membrane was tested for oil separation using five different oil–water mixes, showing high efficiency and good stability and the reusability of the materials developed [6]. Candle soot and polyvinylidene fluoride composites deposited on paper or on aluminium substrates were also evaluated; the materials were considered suitable for such an application due to their superhydrophobicity and superoleophilicity, and their good behaviour was confirmed in tests performed on meshes for both heavy and light oil separation. Such surfaces also proved efficient in protecting stainless steel meshes from concentrated HCl [7]. The specular concept, i.e., developing a superhydrophilic and underwater superoleophobic surface, was applied to obtain oil–water separation with a carbon cloth membrane coated with Cu-doped TiO$_2$ and Ag nanoparticles [8]. Water treatment was the objective of another study, based on melt-blending an ethylene–vinyl alcohol copolymer and polypropylene to produce hydrophilic hollow fibres, which were successfully tested for ink separation [9].

In such applications, durability is a key factor to ensure the industrial applicability of such technologies. Several works, indeed, address this issue, including those previously mentioned in which cycles of oil–water separation were run to monitor performance changes. An assessment of surface durability under mechanical abrasion was performed on dendritic electrodeposited copper surfaces, demonstrating how the precursors affect the final quality and durability of the superhydrophobic effect [10].

Durability was also central in studies related to antibacterial, antiviral and antifouling properties of superhydrophobic materials. In this direction, one study reported the production of electrospun polylactide acid nanofiber filters for face masks: polylactic acid was used as the fibre material, and Manka oil was added to impart antiviral, antibacterial and antifungal activity. Durability was assessed by repeated laundering to ensure reusability [11]. An analogous aim was challenged with the development of anti-wetting coatings on PU foams, produced with fluorine-modified silica, able to prevent fouling from bacteria. Material leaching was then evaluated in both polar and non-polar liquids, representative of liquids typically present in both medical and food industries, which are their target application environments: no release was observed in 14 days of immersion, ensuring the good chemical stability of the coating [12]. Bulk modifications, rather than surface treatments, were evaluated instead in a study dedicated to increasing the durability of concrete through the development of a superhydrophobic concrete mix, which was achieved through the addition of isobutyltriethoxysilane (IBTES) and commercial waterproofing agents. Superhydrophobicity in concrete materials is crucial to increase their service life, as most concrete degradation phenomena are water-related [13].

Finally, two studies delved into more general aspects of self-cleaning. On the one hand, the wetting transition mechanism at the basis of superhydrophobicity was addressed in a review work, beginning with the classical theory of static contact angle and then focusing on the effect of surface morphology through the application of fractal theory and re-entrant geometries [14]. On the other hand, more practical aspects were tackled in a research study aimed at understanding the applicability of the ISO 27448:2009 standard for self-cleaning evaluation to current self-cleaning materials, reviewed in 2020 [15]. Indeed, such a standard was developed for smooth, photocatalytic self-cleaning materials, thus making it inapplicable to the increasingly relevant family of superhydrophobic self-cleaning materials [16].

For further information, readers are encouraged to refer to the complete articles.

Conflicts of Interest: The authors declare no conflicts of interest.

References

1. Diamanti, M.V.; Pedeferri, M.P. Bioinspired Self-cleaning Materials. In *Biotechnologies and Biomimetics for Civil Engineering*; Pacheco Torgal, F., Labrincha, J.A., Diamanti, M.V., Yu, C.-P., Lee, H.-K., Eds.; Springer International Publishing: Berlin/Heidelberg, Germany, 2015; pp. 211–234.
2. Barthlott, W.; Neinhuis, C. Purity of the sacred lotus, or escape from contamination in biological surfaces. *Planta* **1997**, *202*, 1–8. [CrossRef]
3. Barthlott, W.; Mail, M.; Neinhuis, C. Superhydrophobic hierarchically structured surfaces in biology: Evolution, structural principles and biomimetic applications. *Philos. Trans. R. Soc. A* **2016**, *374*, 20160191. [CrossRef]
4. Chabas, A.; Lombardo, T.; Cachier, H.; Pertuisot, M.H.; Oikonomou, K.; Falcone, R.; Verità, M.; Geotti-Bianchini, F. Behaviour of self-cleaning glass in urban atmosphere. *Build. Environ.* **2008**, *43*, 2124–2131. [CrossRef]
5. Watanabe, T.; Nakajima, A.; Wang, R.; Minabe, M.; Koizumi, S.; Fujishima, A.; Hashimoto, K. Photocatalytic activity and photoinduced hydrophilicity of titanium dioxide coated glass. *Thin Solid Films* **1999**, *351*, 260–263. [CrossRef]
6. Zhang, C.; Yang, Y.; Luo, S.; Cheng, C.; Wang, S.; Liu, B. Fabrication of Superhydrophobic Composite Membranes with Honeycomb Porous Structure for Oil/Water Separation. *Coatings* **2022**, *12*, 1698. [CrossRef]
7. Zhang, Y.; Lei, T.; Li, S.; Cai, X.; Hu, Z.; Wu, W.; Lin, T. Candle Soot-Based Electrosprayed Superhydrophobic Coatings for Self-Cleaning, Anti-Corrosion and Oil/Water Separation. *Materials* **2022**, *15*, 5300. [CrossRef] [PubMed]
8. Chen, N.; Sun, K.; Liang, H.; Xu, B.; Wu, S.; Zhang, Q.; Han, Q.; Yang, J.; Lang, J. Novel Engineered Carbon Cloth-Based Self-Cleaning Membrane for High-Efficiency Oil–Water Separation. *Nanomaterials* **2023**, *13*, 624. [CrossRef] [PubMed]
9. Qiu, Z.; He, C. Polypropylene Hollow-Fiber Membrane Made Using the Dissolution-Induced Pores Method. *Membranes* **2022**, *12*, 384. [CrossRef] [PubMed]
10. Akbari, R.; Mohammadizadeh, M.R.; Antonini, C.; Guittard, F.; Darmanin, T. Controlling Morphology and Wettability of Intrinsically Superhydrophobic Copper-Based Surfaces by Electrodeposition. *Coatings* **2022**, *12*, 1260. [CrossRef]
11. Karabulut, F.N.H.; Fomra, D.; Höfler, G.; Chand, N.A.; Beckermann, G.W. Virucidal and Bactericidal Filtration Media from Electrospun Polylactic Acid Nanofibres Capable of Protecting against COVID-19. *Membranes* **2022**, *12*, 571. [CrossRef] [PubMed]
12. Cho, D.; Oh, J.K. Silica Nanoparticle-Infused Omniphobic Polyurethane Foam with Bacterial Anti-Adhesion and Antifouling Properties for Hygiene Purposes. *Nanomaterials* **2023**, *13*, 2035. [CrossRef]
13. Luo, J.; Xu, Y.; Chu, H.; Yang, L.; Song, Z.; Jin, W.; Wang, X.; Xue, Y. Research on the Performance of Superhydrophobic Cement-Based Materials Based on Composite Hydrophobic Agents. *Materials* **2023**, *16*, 6592. [CrossRef] [PubMed]
14. Wang, X.; Fu, C.; Zhang, C.; Qiu, Z.; Wang, B. A Comprehensive Review of Wetting Transition Mechanism on the Surfaces of Microstructures from Theory and Testing Methods. *Materials* **2022**, *15*, 4747. [CrossRef] [PubMed]
15. ISO 27448:2009; Fine ceramics (advanced ceramics, advanced technical ceramics)—Test method for self-cleaning performance of semiconducting photocatalytic materials—Measurement of water contact angle. International Organization for Standardization (ISO): Geneva, Switzerland, 2009.
16. Peeters, H.; Lenaerts, S.; Verbruggen, S.W. Benchmarking the Photocatalytic Self-Cleaning Activity of Industrial and Experimental Materials with ISO 27448:2009. *Materials* **2023**, *16*, 1119. [CrossRef] [PubMed]

Disclaimer/Publisher's Note: The statements, opinions and data contained in all publications are solely those of the individual author(s) and contributor(s) and not of MDPI and/or the editor(s). MDPI and/or the editor(s) disclaim responsibility for any injury to people or property resulting from any ideas, methods, instructions or products referred to in the content.

Article

Research on the Performance of Superhydrophobic Cement-Based Materials Based on Composite Hydrophobic Agents

Jie Luo, Yi Xu *, Hongqiang Chu, Lu Yang, Zijian Song, Weizhun Jin, Xiaowen Wang and Yuan Xue

College of Mechanics and Materials, Hohai University, Nanjing 211100, China; luojiehhu@163.com (J.L.); chq782009@126.com (H.C.); yanglu90@hhu.edu.cn (L.Y.); songzijian@hhu.edu.cn (Z.S.); j_weizhun@163.com (W.J.); 18668911326@163.com (X.W.); 15038950772@163.com (Y.X.)
* Correspondence: xuyihhu@163.com

Abstract: The utilization of a novel monolithic superhydrophobic cement material effectively prevents water infiltration and enhances the longevity of the material. A method for improving superhydrophobic concrete was investigated with the aim of increasing its strength and reducing its cost by compounding superhydrophobic substances with water repellents. The experimental tests encompassed the assessment of the compressive strength, contact angle, and water absorption of the superhydrophobic cementitious materials. The findings demonstrate that an increase in the dosage of isobutyltriethoxysilane (IBTES) progressively enhances the contact angle of the specimen, but significantly diminishes its compressive strength. The contact angle of SIKS mirrors that of SIS3, with a superior compressive strength that is 68% higher. Moreover, superhydrophobicity directly influences the water absorption of cementitious materials, with a more pronounced superhydrophobic effect leading to a lower water absorption rate. The water absorption of cementitious materials is influenced by the combined effect of porosity and superhydrophobicity. Furthermore, FT−IR tests unveil functional mappings, such as -CH_3 which can reduce the surface energy of materials, signifying successful modification with hydrophobic substances.

Keywords: compound hydrophobic agent; contact angle; mechanical properties; water absorption rate; pore structure

Citation: Luo, J.; Xu, Y.; Chu, H.; Yang, L.; Song, Z.; Jin, W.; Wang, X.; Xue, Y. Research on the Performance of Superhydrophobic Cement-Based Materials Based on Composite Hydrophobic Agents. *Materials* 2023, 16, 6592. https://doi.org/10.3390/ma16196592

Academic Editor: Sergei A. Kulinich

Received: 15 September 2023
Revised: 1 October 2023
Accepted: 5 October 2023
Published: 7 October 2023

Copyright: © 2023 by the authors. Licensee MDPI, Basel, Switzerland. This article is an open access article distributed under the terms and conditions of the Creative Commons Attribution (CC BY) license (https://creativecommons.org/licenses/by/4.0/).

1. Introduction

Within the field of engineering, the majority of structures are designed with a lifespan of at least 50 years [1]. Nevertheless, the innate hydrophilic properties of cementitious materials render them vulnerable to the ingress of corrosive agents utilizing water as a medium, consequently compromising their longevity. The hydrophilic characteristic inherent to cement-based materials frequently result in various challenges, including the accumulation of dust [2,3], deterioration due to freeze–thaw cycles [4,5], corrosion caused by chloride salts [6–8], sulfate-induced corrosion [9,10], and the corrosion of reinforcement [11,12]. Structures situated in specific environments, notably those characterized by high humidity, exhibit an elevated susceptibility to premature structural failure. Consequently, it is imperative to minimize the infiltration of moisture into these materials. Presently, researchers are directing their efforts toward the advancement of superhydrophobic surfaces that facilitate unhindered droplet runoff, preventing their adherence to the material's surface [13]. This attribute serves as an effective deterrent against the infiltration of moisture into the inner regions of cementitious materials.

There are two methods for preparing superhydrophobic cement-based materials: surface modification [14–16] and monolithic modification [17]. Surface modification involves the application of coatings or impregnation to create a superhydrophobic film on the material's surface. For example, Vanithakumari et al. [18] prepared superhydrophobic coatings using perfluoro octyl triethoxy silane and hexadecyl trimethoxy silane. Wang et al. [19] prepared superhydrophobic coatings by employing nonfluorinated triethoxyoctylsilane to reduce the material's surface energy, while the combination of microscale diatomaceous earth and sand powders contributed to a hierarchical structure. Wang et al. [20] prepared a superhydrophobic coating using IBTES. However, it is important to note that superhydrophobic coatings are susceptible to environmental damage, which can significantly reduce their working lifespan. Superhydrophobic coatings can easily fail due to cracking of the concrete, aging and the peeling of coatings, etc. [21–23]. Therefore, when considering the long-term protection of concrete buildings, the limited duration of superhydrophobic coatings should be taken into account.

Integral hydrophobic modification involves the use of different hydrophobic substances to lower the surface energy and various methods to create micro and nanostructures, thereby enhancing the material's hydrophobicity. The integral hydrophobic modification of a material for superhydrophobicity can provide a comprehensive range of superhydrophobic properties, not just on the surface. Even if the surface of the material is damaged, the material will still retain its superhydrophobic properties internally, which can extend the effective working time of the material. Several researchers have successfully prepared superhydrophobic concrete using different hydrophobic substances [24–26]. Xu et al. [27] utilized stearic acid and paper sludge ash, ground into a hydrophobic powder with a contact angle of 153°. From an environmental perspective, this powder can replace some of the silicate cement. Wang et al. [28] prepared hydrophobic concrete by using stearic acid and further polishing the concrete surface with sandpaper, which results in superhydrophobic concrete with a contact angle of 153.7°.

Despite its potential advantages, it is vital to acknowledge that integral hydrophobic modification for achieving superhydrophobicity in materials may lead to a notable reduction in their mechanical properties. Wang et al. [29] conducted research demonstrating that superhydrophobic modification resulted in a substantial reduction of 30.9% in compressive strength and 18.1% in flexural strength compared with unmodified materials. Dong et al. [30] used oil-in-water suspension emulsions to prepare superhydrophobic concrete, and the compressive strength decreased significantly from 8.7 MPa to 1.4 MPa as the oil/water volume ratio increased from 1:1 to 4:1. Furthermore, it is worth noting that some of hydrophobic agents employed in these methodologies are environmentally unfriendly, containing polluting elements [18] such as fluorine (F). Additionally, these agents can significantly reduce the strength of cementitious materials, increase costs [31] and affect their practical application in engineering.

This study aims to prepare an environmentally friendly superhydrophobic cementitious material by utilizing a novel composite hydrophobic agent. The compound hydrophobic agent significantly improves the superhydrophobicity of the material, has a low effect on strength, and can reduce costs. When the contact angles are similar, the composite hydrophobic agent has a higher compressive strength and lower cost compared with IBTES. Additionally, this study investigates the influence of IBTES and the composite hydrophobic agent on the contact angle, mechanical properties and water absorption of the cementitious material. X-ray diffractometer (XRD) and Fourier transform infrared spectroscopy (FT−IR) are employed to explore the mechanisms underlying the superhydrophobic properties of the cementitious material, while mercury intrusion porosimetry (MIP) is used to analyze the pore structure of the superhydrophobic cementitious material.

2. Materials and Methods
2.1. Materials

Portland cement (P·II 42.5 compliant with GB 175-2007 and equivalent to CEM I 42.5) from Nanjing Hailu Cement Co., Ltd., Nanjing, China. was used. Nano silica (NS; BET: 200 m^2/g, average diameter: 13 nm) was obtained from the Beimo company (Jiaxing, China). Silica fume (SF; Model 951) was obtained from the Elkem company, Oslo, Norway. Isobutyltriethoxysilane (IBTES) was obtained from RHΛWN, Shanghai, China. Polycarboxylic superplasticizer (water reducing efficiency: 30%) and waterproofing agent (model number: KLJ®-VI) were obtained from Sobute New Materials Co., Ltd., Nanjing, China.

2.2. Mix Proportion

The test reference ratio was a water/cement ratio (W/C) of 0.5 and a cement/sand ratio (C/S) of 1:2. The mix proportions are shown in Table 1. The dosages in the table are all mass fractions. All other substances were added as a percentage of the dosage of cement. SIS1, SIS2 and SIS3 groups were used to study the effect of different dosage of IBTES on cement-based materials. To reduce costs, the SIKS were considered compound hydrophobic agents. SIKS were doped with silane. For the comparison test, the SKS group contained only KLJ. PC1 and PC2 served as blank control groups for the study. In PC2, silica fume and silica nanoparticles were used as dopants, while PC1 remained free of any dosage agent.

Table 1. Matching ratio design (mass fraction %).

Sample	Cement/g	W/C	C/S	NS	SF	IBTES	KLJ
PC1	600	0.5	0.5	-	-	-	-
PC2	600	0.5	0.5	2%	6%	-	-
SIS1	600	0.5	0.5	2%	6%	4%	-
SIS2	600	0.5	0.5	2%	6%	6%	-
SIS3	600	0.5	0.5	2%	6%	8%	-
SIKS	600	0.5	0.5	2%	6%	4%	8%
SKS	600	0.5	0.5	2%	6%	-	8%

2.3. Superhydrophobic Concrete Preparation

The process of preparing superhydrophobic cement-based materials is shown in Figure 1. The mass of each raw material was weighed according to the test ratio, and the mixed solution was prepared by first dispersing the NS in water and stirring for 30 s. Then, the hydrophobic substance (IBTES, KLJ) was added to a coagulation solution of NS and water. The above solution was dispersed via ultrasonication at 40 °C for 30 min to prepare a superhydrophobic solution. When forming, cement, sand and SF were put into the mixer in turn. After stirring for 1 min, the super hydrophobic solution was then added and stirred for another 4 min to obtain fresh superhydrophobic mortar mixture. Then, the fresh superhydrophobic mortar mixture was poured into the mold, a nylon net was attached to the surface of the mold and the mold was demolded after 1 d and put into the maintenance room (temperature of 20 °C and humidity of 80%) for 28 d.

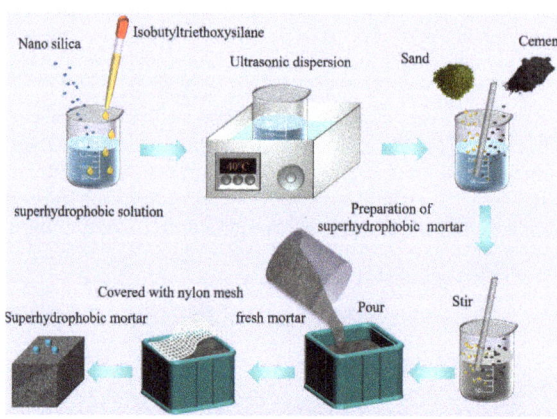

Figure 1. The schematic diagram of the fabrication process of superhydrophobic mortar.

2.4. Test Methods

2.4.1. Wettability Test

Contact angle (WCA) and contact angle hysteresis (CAH) were performed with a contact angle meter (DSA30, Kruess, Hamburg, Germany) at ambient temperature to characterize the superhydrophobic property. The volumes of probing liquids were approximately 5 μL for the contact angle measurement. Each contact angle reported was an average value of five independent measurements on different spots. The droplet images of the specimen surface were taken with the digital camera and macro lens that come with the contact angle measuring instrument.

2.4.2. Compressive Strength Test

The compressive strength test was conducted in accordance with the Chinese national standard GB/T 17671-2021. All samples were cured for 28 d before testing. The cubic specimens (40 mm × 40 mm × 40 mm) were prepared to measure the compressive strength. The final result was determined by averaging the values obtained from three specimens.

2.4.3. Waterproofing Ability Test

The water absorption test was performed with reference to the Chinese national specification JGJT 70-2009. The test was performed using repeated samples on ordinary and superhydrophobic cement-based materials, and after the age reached 28 d, these samples were put into an oven at 80 °C, dried to constant weight and then were put into water, and the distance from the surface of the specimen to the water surface was kept at 3 cm throughout the process. The weight of the specimen was weighed at regular intervals. The water absorption rate calculation formula is shown in Equation (1). Three specimens were taken from each group and the average value was taken after measurement, and the trend of the water absorption rate of each group was observed with time.

$$W = \frac{m_1 - m_0}{m_0} \times 100 \tag{1}$$

where W is the water absorption of test pieces, m_1 is the weight of test pieces at different times, and m_0 is the initial weight of test pieces after drying.

2.4.4. Characterization

The surface morphology of the samples was analyzed using a scanning electron microscope (SEM) model ZEISS Sigma 300, Carl Zeiss Microscopy GmbH, Oberkochen, Germany and their chemical compositions were investigated using energy-dispersive spectroscopy (EDS, ZEISS Sigma 300, Carl Zeiss Microscopy GmbH, Oberkochen, Germany). XRD analysis was used to characterize the specific composition of the sample product. The identification of phase compositions was conducted using an X-ray diffractometer (XRD, Rigaku-D/max 2200pc, Rigaku Corporation, Tokyo, Japan) with a Cu Ka (k = 1.54 Å) incident radiation. The 2θ scanning range was from 5° to 90° with a scanning speed of 2 °/min. The pore structure of the hardened cement paste was analyzed using mercury intrusion porosimetry (MIP, MicromeritiPC1 AutoPore V 9620, Micromeritics, Norcross, GA, USA). The analysis of functional groups was undertaken using Fourier transform infrared spectroscopy (FT−IR, Thermo Scientific Nicolet iS20, Thermo Fisher Scientific, Waltham, MA, USA).

3. Results and Discussion

3.1. Wettability

Figure 2 illustrates the morphology of the water droplets on a superhydrophobic surface. This indicates that the superhydrophobic surface was successfully prepared. The effect of the hydrophobic agent type and dosage on the contact angle is presented in Figure 3a. Within the single hydrophobic group (SIS1, SIS2, SIS3), evidence is presented that as the dosage of IBTES increases, the contact angle progressively rises, reaching a remarkable 158.3°. The plain concrete (PC1 and PC2) without hydrophobicity had a contact angle of less than 90°. The PC1 and PC2 exhibited hydrophilicity. The KLJ was able to modify the hydrophobicity of cement-based materials. But the contact angle of SKS was only 111.1°, and did not display superhydrophobicity. The composite hydrophobic agents prepared by using the KLJ and IBTES were effective in the superhydrophobic modification of cement-based materials.

Figure 2. Morphology of the water droplet on superhydrophobic surfaces.

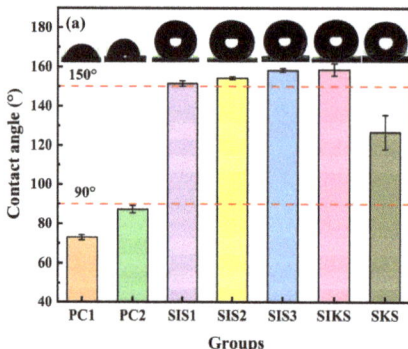
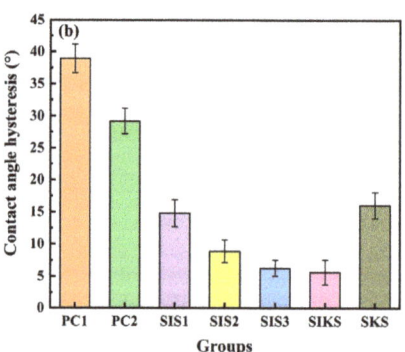

Figure 3. Effect of different hydrophobic agents on (a) contact angle and (b) contact angle hysteresis.

Upon comparing the composite hydrophobic agent with the single hydrophobic agent (SIKS compared with SIS1, SIS2, SIS3), it is evident that the IBTES dosage in the SIKS group was the same as that in SIS1, but the contact angle increased from 151.5° to 158.6° (Figure 3a). These results indicate that the addition of KLJ effectively enhanced the superhydrophobic modification when combined with IBTES. The superhydrophobicity of the SIKS and SIS3 were similar, with contact angles of approximately 158°. However, the SIS3 group contained double the amount of IBTES compared with the SIKS group, and this excessive IBTES content adversely affected the mechanical properties of cementitious materials. The contact angles of PC1 and PC2 were less than 90°, while the contact angles of the other groups were more than 90°. This indicates that hydrophobic substances can effectively convert the gelling material from hydrophilic to hydrophobic. The IBTES and KLJ reduced the surface energy of the material and modified the material. The contact angle hysteresis of SIS1, SIS2 and SIS3 decreased gradually (Figure 3b). PC1 and PC2 were not superhydrophobically modified, with a contact angle hysteresis of 29° or more. SIKS had a contact angle hysteresis of less than 10°. SKS was only hydrophobically modified and had a contact angle hysteresis of more than 10°.

Figure 4 illustrates the wettability of various specimen surfaces. The superhydrophobic surface exhibited a water droplet in spherical form that swiftly rolled off, leaving no droplets behind. On the other hand, the hydrophobic surface showed a transition. As the water droplet increased in size, it gradually became oval and slowly slid off, usually leaving a mark on the surface. In contrast, the hydrophilic surface behaved differently. With an increasing volume of the water droplet, the surface gradually became wetted and did not facilitate the formation of a rolling water droplet.

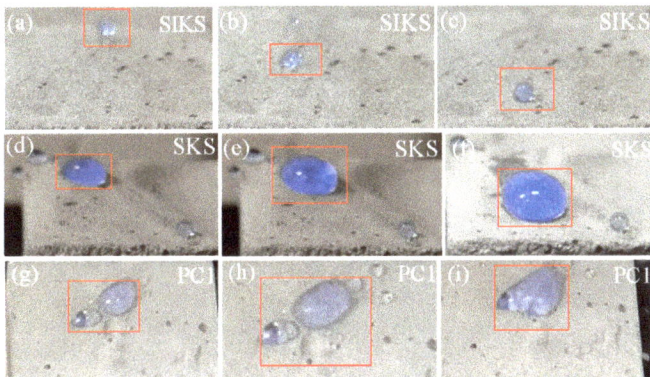

Figure 4. Water droplet wetting diagram on specimen surface: (**a**–**c**) superhydrophobic; (**d**–**f**) hydrophobic; (**g**–**i**) hydrophilic.

The dynamic bouncing processes of water droplets on the surface of different specimens are shown in Figure 5. A typical water bouncing process could be obtained on the surface of the superhydrophobic mortar, where the water droplets can leave the surface completely without leaving any residue. On PC1, PC2 and SKS surfaces, water droplets could not leave the surface. And water droplets were adsorbed by the surface. The SKS water bead bounce was higher than PC1 and PC2. This suggests that hydrophobic surfaces have an effect on dynamic bouncing process. When a water droplet touched the surface of PC1 and PC2, the droplet was adsorbed directly onto the surface and detached from the syringe (Figure 6). Water droplets were adsorbed on the surfaces of SIS1, SIS2 and SIS3 and were dislodged from the syringe. This suggests that the Cassie–Baxter state [32] is not stable enough. In SIKS, the water droplet did not adhere to the surface after contacting it. Finally, the water droplet adhered to the hydrophobic tip of the syringe and left the superhydrophobic surface when the syringe rose from the surface. This suggests that the Cassie–Baxter state is stable under external forces [33].

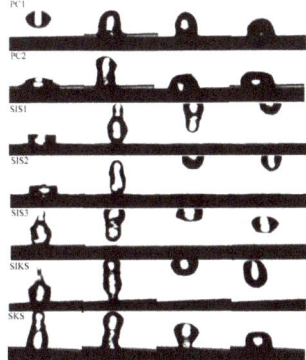

Figure 5. Dynamic bouncing processes of different groups.

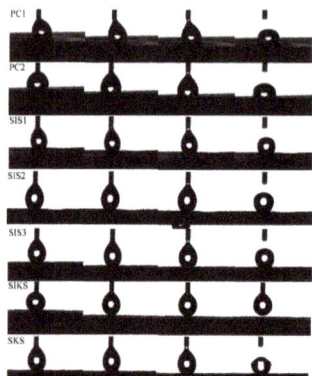

Figure 6. Water droplets touching different surfaces.

3.2. Compressive Strength

The compressive properties of different hydrophobic agents on cement-based materials are depicted in Figure 7. The PC2 showed a higher compressive strength compared with the PC1 because PC2 is doped with nano-silica and silica fume. With an increase in IBTES dosage, the compressive strength of each SIS group gradually decreased. In comparison to the PC1, the addition of IBTES reduced the compressive strength of SIS1, SIS2 and SIS3 by 4.8%, 20.3% and 35.0%, respectively. Similarly, when contrasted with the PC2, the compressive strength of SIS1, SIS2 and SIS3 exhibited reductions of 21.9%, 33.1% and 45.4%, respectively. The compressive strength of the singly doped group declined sharply as the amount of IBTES increased, and when the contact angle reached 158° or higher the compressive strength decreased by more than 35%. In order to enhance the compressive strength without affecting the superhydrophobicity, a compound hydrophobic agent was used instead of a single hydrophobic agent. Compared with PC1, the compressive strength of SIKS improved by 9.3%. The compound hydrophobic agent could improve the compressive strength while maintaining a contact angle above 155°. Compared with SIS3, the compressive strength of SIKS increased by 68.0%. The contact angles of SIS3 and SIKS were close, but the compressive strength of SIKS was higher than that of SIS3. Although the compressive strength of SKS was higher than that of SIKS, the contact angle of SIKS was lower than 150°, which was not superhydrophobic. This suggests that the effect of IBTES on superhydrophobicity was more important. The KLJ could only hydrophobically modify mortar, but had a positive effect on strength.

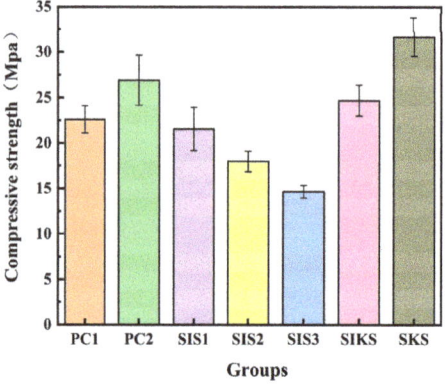

Figure 7. Effect of different hydrophobic agents on compressive strength.

The KLJ exhibited greater strength compared with both PC1 and PC2 due to its ability to react with cement hydration products, forming insoluble crystals that block concrete capillaries and enhance concrete density [34]. The KLJ also refined the capillary structure of concrete, increased its density, and mitigated the adverse effects of IBTES on strength. The SIKS was enriched with nanomaterials, yielding a positive effect on strength enhancement [35]. Meanwhile, the substitution of a portion of IBTES with KLJ further bolstered the material's strength. Nanoparticles and silica fume [36] facilitated cement hydration, consequently enhancing the material's compressive strength while mitigating the detrimental effects of hydrophobic agents. Nano-silica and silica fume possessed fine particles capable of filling mortar pores and micro-cracks, thereby increasing the mortar compactness and density. Additionally, nano-silica and silica fume can serve as crystalline nuclei, fostering hydration reactions during cement gel formation [37]. The hydrophobic substance affected cement hydration and reduced the strength of mortar. The IBTES does not contain element F and is environmentally friendly, but is expensive. The KLJ replaced some of the IBTES and can effectively reduce costs. While the contact angle of SIKS closely resembles that of SIS3, its compressive strength significantly surpassed that of SIS3. This observation implies that the composite hydrophobic agent can achieve the superhydrophobic modification of cementitious material with minimal effects on compressive strength.

3.3. Water Absorption Rate

The water absorption rates of the specimens in different groups are shown in Figure 8, and the water absorption rates of the other superhydrophobic test groups were lower compared with PC1 and PC2. Initially, the water absorption rates rose rapidly in all groups. As the water absorption of the specimen reached saturation, the rate of increase of water absorption gradually slowed down and eventually stabilized at a constant value. In the first hour, the PC1 group had a water absorption rate of 6.2%. The water absorption of the PC2 with nanomaterials was much lower, at 3.7%, a reduction of about 40%. After 200 h, PC1 and PC2 had water absorptions of 7.9% and 7.2%, respectively. In the single dosage group (SIS1, SIS2, SIS3), as the amount of IBTES increased, superhydrophobicity gradually increased and the water absorption decreased significantly. In the first hour, the highest water absorption rate among these groups was only 0.4%, much lower than PC1 and PC2. Following 200 h, the water absorption rates in the single-doped groups (SIS1, SIS2, SIS3) were 2.9%, 2.9% and 2.8%, respectively. These values signify a noteworthy decline of 62.9%, 64.0% and 64.4% in comparison with PC1. Superhydrophobic cement-based materials effectively hindered moisture from infiltrating the material's interior, which resulted in a decrease in water absorption rates as superhydrophobicity increased. The SKS exhibited a 5.5% water absorption rate after 200 h, marking a reduction of 29.9% compared with PC1. The SKS had a contact angle of less than 150°, which was not superhydrophobic and offered limited waterproofing protection inside the material. As a result, it had a higher water absorption rate than the other superhydrophobic groups. In contrast, the compound hydrophobic agent (SIKS) yielded promising outcomes and the water absorption rate was consistently below 1% within the first hour. Over 200 h, the water absorption rate for SIKS remained below 4%, indicating a substantial 59.3% reduction compared with PC1.

The water absorption rate of PC2 is lower than that of PC1. This difference is due to the incorporation of nanomaterials in the PC2, which results in material compaction and a reduction in water absorption. This suggests that filling the pores of the cementitious material with nanomaterials has a certain effect on reducing water absorption, although it is not very pronounced. The reduction in water absorption can be attributed to two factors: the improvement in material densification [38] and the hydrophobic modification of the material [39]. The water absorption of the test group mixed with hydrophobic material was significantly lower, at only 4.4%, while the water absorption of the PC1 group was 7.9%. Achieving a significant reduction in water absorption and improving material durability required the transformation of hydrophilic materials into superhydrophobic ones. This transformation fundamentally prevented water infiltration into the material. The contact angle of the SKS was below 150°. This suggests that the single-doped hydrophobic chemical pore bolus had a lesser effect on material modification when compared with other test groups. As a result, the water absorption rate in the SKS group exceeded that of other superhydrophobic test groups. This demonstrated a clear correlation between the superhydrophobic effectiveness of modification and water absorption. Enhanced superhydrophobicity led to a decreased rate of water infiltration into the material, which resulted in reduced water absorption.

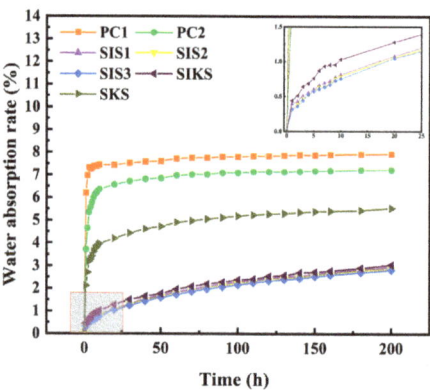

Figure 8. Variation of water absorption rate of each group with time.

3.4. SEM Analysis

Figure 9 shows the SEM images of the superhydrophobic concrete at different magnifications. At the 3 μm scale, the structure of PC1 was looser, with many pores and C-S-H cross-linking each other (Figure 9a). When the magnification was 30 KX, the microscopic morphology of the PC1 surface was that of inter-crosslinked C-S-H. Flaky C-S-H gels were generated in PC1, which indicates a high degree of hydration. The morphologies of SIS1, SIS2 and SIS3 are relatively similar, all being C-S-H gels crosslinked with each other and featuring numerous nano-silica particles on their surfaces. These silica nanoparticles created rough structures that significantly enhanced the superhydrophobicity of cement-based materials. Roughness plays a crucial role in the development of superhydrophobic surfaces. The internal structure of the superhydrophobic specimen revealed the presence of numerous micro- and nano-scale bumps. These rough structures facilitated the capture and retention of air layers, allowing the water droplet to form large contact angles and effortlessly roll off inclined surfaces [40,41]. As the amount of IBTES dosage increased, there was a gradual increase in the number of micro-nano structures within the material. The micro-nano structures further enhanced the superhydrophobic modification effect. The SIS1 exhibited lower porosity, the SIS2 formed a lamellar C-S-H gel with a higher degree of hydration, and the SIS3 possessed a looser structure. These characteristics aligned with their respective compressive strength results. With the increase in IBTES dosage, the structure gradually became looser, and the pore space gradually expanded. The presence of hydrophobic substances in the cement paste had a notable effect. The hydrophobic substance reduced the fluidity of the cement paste, which made it challenging for water to permeate between the cement particles and hydrophobic substances and affected the hydration reaction of the cement. The SIKS produced a denser structure of layered C-S-H gels (Figure 9i). This was consistent with its higher compressive strength results. The amount of IBTES in SIKS was smaller, so the structure of the SIKS was denser compared with SIS3. This is consistent with the previous results of compressive strength. The SKS has many fine particles within it, which fills the pores and makes the structure denser. Figure 10 shows the elemental analysis of SIKS (Figure 9i, EDS1) and SKS (Figure 9l, EDS2). The SIKS and SKS contained basically the same elements, mainly elements contained in the hydration products of cement, such as Ca, Si, etc. The Ca/Si of SIKS was about 0.5, within the typical range of C–S–H gel [42]. The Ca/Si of SIKS was about 0.1. This shows that it is not a C-S-H gel. The C content of SIKS was higher than the SKS. This suggests that there was more hydrophobic material in the SIKS hydration product.

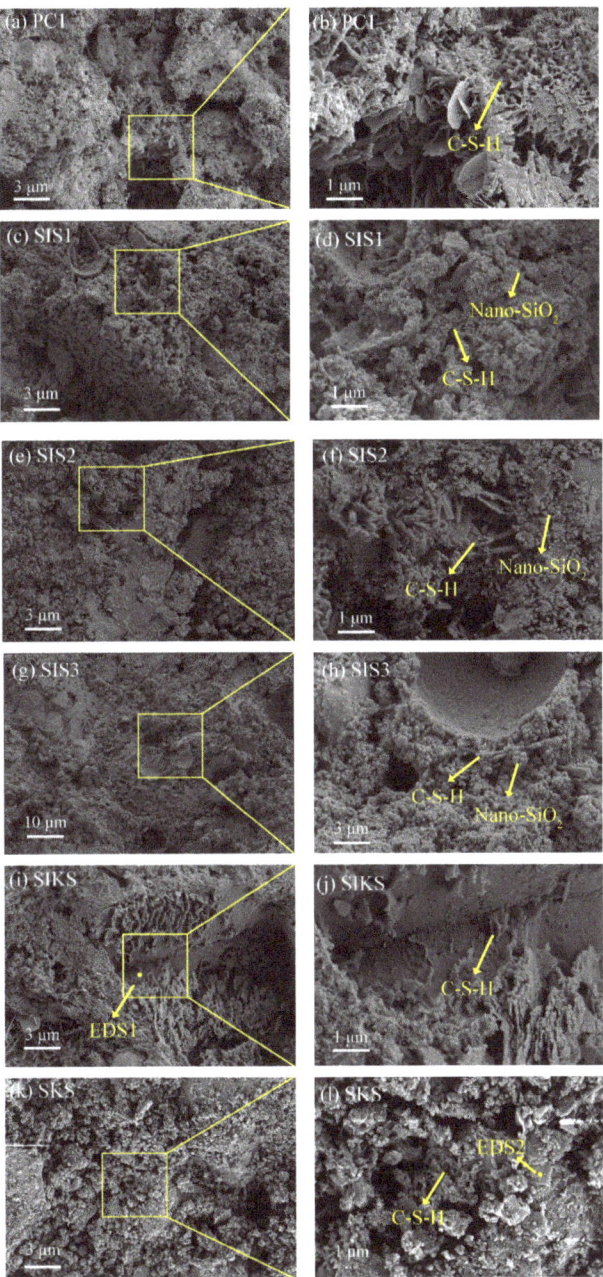

Figure 9. Micro morphology of different groups.

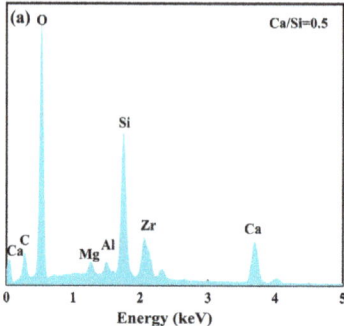
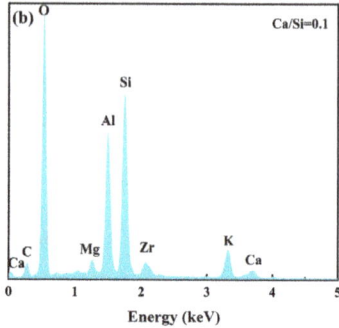

Figure 10. EDS spectrum of (a) SIKS and (b) SKS.

3.5. XRD Analysis

XRD is used to analyze the specific chemical composition of the material. The crystal structures of the different specimens were detected using the XRD technique to determine their main components. The main components of different specimens were similar, both containing Ca(OH)$_2$, AFt, SiO$_2$, etc. (Figure 11). This indicates that IBTES and KLJ did not change the chemical composition of the specimens. The highest peak in the position of 26.6° was in accordance with quartz sand (JCPDS No. 46–1045). Peaks at 18.0° and 34.2° were in line with calcium hydroxide (JCPDS No. 04–0733) and the peak at 9.0° was in keeping with AFt (JCPDS No. 41–1451). This was similar to the results of reference [43]. The peaks of Ca(OH)$_2$ in SKS were much higher than the other groups, indicating that IBTES affected the hydration of cement to produce C-S-H gels. Different hydrophobic substances do not produce new peaks, which indicates that hydrophobic substances do not contribute much to the crystal structure [44].

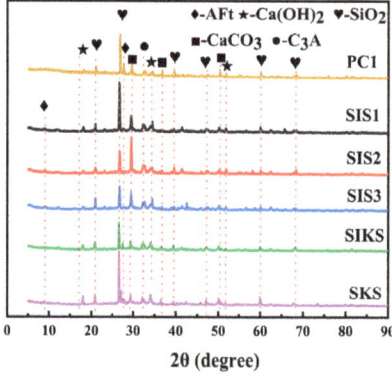

Figure 11. XRD patterns of superhydrophobic concrete prepared with different hydrophobic agent types.

3.6. FT−IR Analysis

In addition to the surface morphology, the chemical composition also affects the wettability of the material surface. In order to characterize the effect of superhydrophobic modification, an analysis of functional groups was performed using FT−IR. Figure 12 shows the FT−IR spectra of the different specimens. A stretching vibration peak of -OH of water was present at 3437 cm^{-1} for each group. In superhydrophobic concrete, the peak observed at 2945 cm^{-1} may be the stretching vibration of C-H in hydrophobic substances. Distinct peaks were observed near 2922 cm^{-1} and 2850 cm^{-1} in the SIS1, which confirms the presence of -CH$_2$ and -CH$_3$ functional groups. The stretching vibration of Si-O was observed at 1111 cm^{-1} for SIS1, SIS2, SIS3, SIKS and SKS. The presence of hydrophobic functional groups within the material was confirmed by the observed absorption peaks. However, PC1 did not show hydrophobic functional groups at the same position. These substances were effectively integrated into the cementitious materials through a sequence of chemical processes involving hydrolysis, dehydration and condensation [45], which resulted in the development of superhydrophobic properties within the cementitious materials.

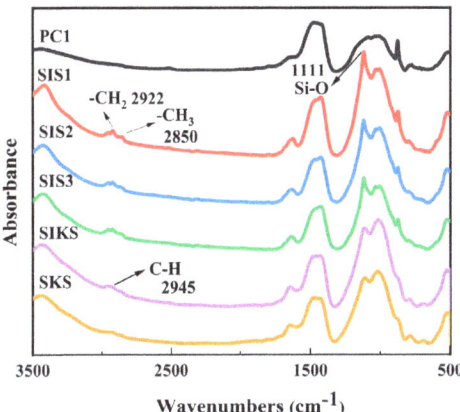

Figure 12. FT−IR patterns of superhydrophobic concrete prepared with different hydrophobic agent types.

The mechanism of superhydrophobicity is depicted in Figure 13. The siloxane group [46] undergoes hydrolysis to form a silanol group. The silanol group readily undergoes a dehydration condensation reaction with the hydroxyl group. This process facilitates the adhesion of low surface energy substances to the material's surface, effectively reducing the solid surface energy. According to the Cassie–Baxter theoretical model [32], an increase in solid surface roughness leads to an increase in the contact angle on a hydrophobic surface. Merely reducing the surface energy of solids can only modify hydrophilic materials to hydrophobic materials; it cannot achieve a superhydrophobic effect. Therefore, a rough structure was introduced on the material surface using a nylon net. This dual action of low surface energy and rough structure modified the material into a superhydrophobic state. The SKS group had a higher compressive strength than the other test groups, and its contact angle was also greater than 90°. This indicates that the KLJ not only reduced the surface energy of the material, but also promoted the hydration of the cement and increased the compressive strength of the material. Composite hydrophobic agents prepared from KLJ and IBTES had a better effect on superhydrophobicity and compressive strength. The spectral analysis of SIKS and SKS contained C (Figure 10). Spectral analysis by SIKS shows that the C-S-H gels contained C. This suggests that IBTES hydrolyzed and cross-linked with the C-S-H gels.

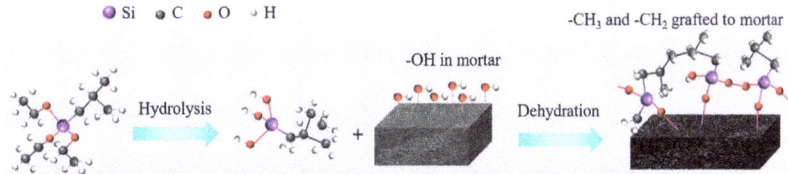

Figure 13. Superhydrophobic modification mechanism of hydrophobic agents.

3.7. MIP Analysis

The pore parameters of the superhydrophobic cementitious materials prepared with different hydrophobic agents are shown in Figure 14. The pore volumes of the SIS1, SIS2, SIS3, SIKS and SKS groups were 0.1124, 0.1129, 0.1258, 0.1000 and 0.1081 mL/g, respectively (Figure 14a). The pore volume in the compound dosage group (SIKS) was lower than that in the single dosage group (SIS1, SIS2, SIS3). This suggests that the KLJ contributes to pore compaction and structural densification. The pore volume steadily rose within the single dosage group (SIS1, SIS2, SIS3) as the IBTES dosage increased. This phenomenon may be attributed to the influence of IBTES on hydration, impeding the compact growth of C-S-H. The difference in pore volumes between the groups was not significant and indicated that the nanomaterials filled the pores and enhanced material compactness. Mercury compression measurements show that the SIKS and SKS groups had the lowest porosity, while the SIS3 group exhibited the lowest water absorption. This indicates that porosity had an impact on water absorption, but it is important to recognize that superhydrophobicity also exerts an influence on water absorption. The porosity and superhydrophobicity play a combined role in determining water absorption in cementitious materials.

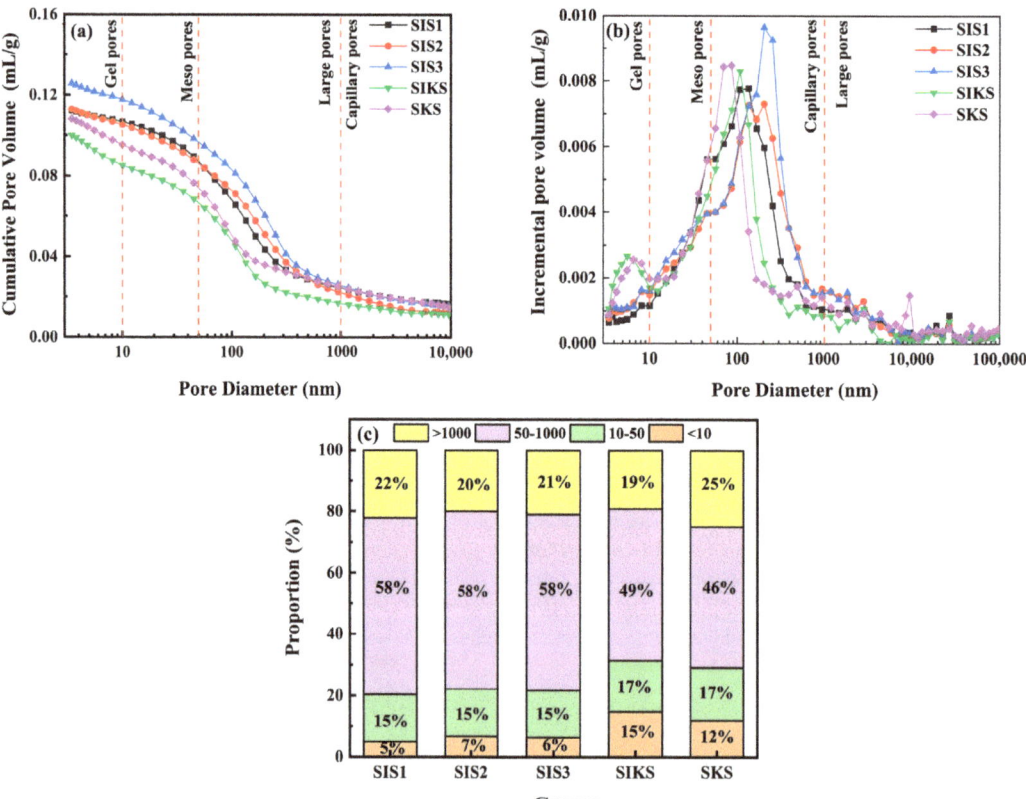

Figure 14. MIP test for pore size distribution: (**a**) cumulative pore volume; (**b**) pore structure distribution; (**c**) pore diameter distribution.

Related studies have shown that pores with a pore size below 10 nm are gel pores, capillary pores with a pore size between 10 and 50 nm are mesopores, and pores with a pore size between 50 and 10,000 nm are referred to as macropores [47]. It is generally believed that the more pores less than 50 nm, and the fewer the pores larger than 100 nm, the better the performance of concrete. The pore distribution maps of different specimens demonstrated percentages of different pore sizes by 10 nm, 50 nm and 1000 nm (Figure 14b). The percentages of macropores in the SIS1, SIS2, SIS3, SIKS and SKS groups were 79.61%, 77.92%, 78.32%, 68.56% and 70.83%, respectively. The total porosity in the SIS1, SIS2, SIS, SIKS and SKS groups are shown in Table 2. The larger the percentage of large pores, the more unfavorable it is for cementitious materials. Nano-silica and silica fume can densify the pore space, but the hydrophobic agent will be wrapped on the surface of cement particles and affect the hydration of cement. The hydrophobic agent affects the gelling of hydration products and hinders the connection of cement particles to form a continuous network structure. This, in turn, causes loose hydration products to impact the pore space within the structure. The volcanic ash effect generates more C-S-H gels, which leads to an increase in the number of mesopores and gel pores in the cementite. Another important pore parameter is the threshold pore size, which is closely related to the connectivity of the pores and the permeability of the cementitious material. The threshold apertures for the SIS1, SIS2, SIS3, SIKS and SKS groups are shown in Table 2. The SKS had the smallest threshold apertures. This indicates that the pore connectivity of the SKS group was lower and the pore structure was denser [48,49]. This was consistent with the results of the SKS micro-morphology. The SEM picture of SKS had a dense structure with a smaller pore structure (Figure 9k).

Table 2. Total porosity and the most probable pore diameter for different samples.

	SIS1	SIS2	SIS3	SIKS	SKS
Total porosity (%)	21.59	21.58	23.33	19.14	20.99
The most probable pore diameter (nm)	136.14	205.17	205.08	108.22	86.37

4. Conclusions

The effects of the composite hydrophobic agent and IBTES on the contact angle, mechanical properties and water absorption of cementitious materials were investigated. The mechanism related to the superhydrophobic properties of cementitious materials was investigated using XRD, FT−IR and other microscopic tests, and the pore structure of superhydrophobic cementitious materials was studied using MIP, and the following conclusions can be drawn:

(1) With the increase in IBTES dosage, the superhydrophobicity of the specimen gradually increased, but the strength decreased significantly. The compressive strength of composite hydrophobic agent was higher than that of IBTES when the contact angle was close, and the cost was lower. By using the lower-cost KLJ instead of higher-cost IBTES, the cost can be reduced and engineering applications will be facilitated. Hydrophobic substances have a detrimental effect on the compressive strength of cementitious materials. In the SKS group, the superior effect of compressive strength is attributed to the promotion of the hydration reaction by the nanomaterials, which outweighs the inhibition of the hydration reaction by the hydrophobic substances.

(2) The water absorption rates of SIS1, SIS2, SIS3 and SIKS groups after 200 h were less than 3.1%. Compared with PC1, the water absorption rates were reduced by more than 59%. The water absorption is related to two factors: porosity and superhydrophobicity. The effect of superhydrophobic modification is directly related to the water absorption rate. The better the modification effect, the more difficult it is for water to enter the interior of the material, and the lower the water absorption rate of the material.

(3) FT−IR analysis showed that hydrophobic functional groups were successfully grafted onto the material. The incorporation of IBTES increased the porosity of the material, while the LJK facilitated the dense structural pores and reduced the porosity.

Author Contributions: Conceptualization, J.L. and Y.X. (Yi Xu); methodology, Y.X. (Yi Xu); software, X.W.; validation, J.L., Y.X. (Yi Xu) and Z.S.; formal analysis, H.C.; investigation, H.C.; resources, L.Y.; data curation, J.L.; writing—original draft preparation, J.L.; writing—review and editing, L.Y.; visualization, W.J.; supervision, W.J. and Y.X. (Yuan Xue); project administration, X.W. and Y.X. (Yuan Xue); funding acquisition, Y.X. (Yi Xu) and Z.S. All authors have read and agreed to the published version of the manuscript.

Funding: We gratefully acknowledge the following funds for their support of this study: the National Natural Science Foundation of China (52178203 and 52071130) and the Natural Science Foundation of Jiangsu Province (BK20211204).

Institutional Review Board Statement: Not applicable.

Informed Consent Statement: Not applicable.

Data Availability Statement: Not applicable.

Conflicts of Interest: The authors declare no conflict of interest.

References

1. Tang, S.W.; Yao, Y.; Andrade, C.; Zhong, J. Recent durability studies on concrete structure. *Cem. Concr. Res.* **2015**, *78*, 143–154.
2. Shen, W.G.; Zhang, C.; Li, Q.; Cao, L.; Ye, J.Y. Preparation of titanium dioxide nano particle modified photocatalytic self-cleaning concrete. *J. Clean. Prod.* **2015**, *87*, 762–765. [CrossRef]
3. Visali, C.; Priya, A.K.; Dharmaraj, R. Utilization of ecofriendly self-cleaning concrete using zinc oxide and polypropylene fibre. *Mater. Today Proc.* **2021**, *37*, 1083–1086.
4. Tang, H.; Yang, Y.; Li, H.; Xiao, L.F.; Ge, Y.F. Effects of chloride salt erosion and freeze–thaw cycle on interface shear behavior between ordinary concrete and self-compacting concrete. *Structures* **2023**, *56*, 104990.
5. Ren, J.G.; Lai, Y.M. Study on the durability and failure mechanism of concrete modified with nanoparticles and polypropylene fiber under freeze-thaw cycles and sulfate attack. *Cold Reg. Sci. Technol.* **2021**, *188*, 103301.
6. Xu, W.Q.; Li, Y.; Li, H.Z.; Wang, K.; Zhang, C.P.; Jiang, Y.B. Corrosion mechanism and damage characteristic of steel fiber concrete under the effect of stray current and salt solution. *Constr. Build. Mater.* **2022**, *314*, 125618.
7. Jain, A.; Gencturk, B. Multiphysics and Multiscale Modeling of Coupled Transport of Chloride Ions in Concrete. *Materials* **2021**, *14*, 885.
8. Zhang, F.; Hu, Z.; Wei, F.; Wen, X.; Li, X.G.; Dai, L.; Liu, L. Study on concrete deterioration in different NaCl-Na$_2$SO$_4$ solutions and the mechanism of Cl$^-$ diffusion. *Materials* **2021**, *14*, 5054.
9. Selyaev, V.; Neverov, V.; Selyaev, P.; Sorokin, E.; Yudina, O. Predicting the Durability of Concrete Structures, including Sulfate Corrosion of Concrete. *Mag. Civ. Eng.* **2014**, *45*, 41–52.
10. Li, T.Y.; Wang, W.K.; Li, Y.T.; Bao, T.F.; Zhao, M.F.; Shen, X.X.; Ni, L.; Ma, Q.L.; Tian, H.W. Corrosion Failure Mechanism of Ultra-high-performance Concretes Prepared with Sea Water and Sea Sand in an Artificial Sea Water Containing Sulfate. *J. Chin. Soc. Corros. Prot.* **2023**, *43*, 1101–1110.
11. Green, W.K. Steel reinforcement corrosion in concrete—An overview of some fundamentals. *Corros. Eng. Sci. Technol.* **2020**, *55*, 289–302.
12. Chen, D.; Mahadevan, S. Chloride-induced reinforcement corrosion and concrete cracking simulation. *Cem. Concr. Compos.* **2008**, *30*, 227–238.
13. Li, Z.J.; Guo, Z.G. Self-healing system of superhydrophobic surfaces inspired from and beyond nature. *Nanoscale* **2023**, *15*, 1493–1512. [PubMed]
14. Dong, W.K.; Li, W.G.; Sun, Z.H.; Ibrahim, I.; Sheng, D.C. Intrinsic graphene/cement-based sensors with piezoresistivity and superhydrophobicity capacities for smart concrete infrastructure. *Autom. Constr.* **2022**, *133*, 103983.
15. Jiang, T.; Fan, W.X.; Wang, F.J. Long-lasting self-cleaning daytime radiative cooling paint for building. *Colloids Surf. A Physicochem. Eng. Asp.* **2023**, *666*, 131296.
16. Yu, X.F.; Shi, X.T.; Xue, F.X.; Bai, W.X.; Li, Y.W.; Liu, Y.H.; Feng, L.B. SiO$_2$ nanoparticle-containing superhydrophobic materials with enhanced durability via facile and scalable spray method. *Colloids Surf. A Physicochem. Eng. Asp.* **2021**, *626*, 127014.
17. Zhao, Y.Y.; Lei, L.L.; Wang, Q.; Li, X.D. Study of superhydrophobic concrete with integral superhydrophobicity and anti-corrosion property. *Case Stud. Constr. Mater.* **2023**, *18*, e01899.
18. Vanithakumari, S.C.; Thinaharan, C.; Philip, J. Polyvinylidene Difluoride-SiO$_2$-Silane Based Superhydrophobic Coatings on Carbon Steel with Enhanced Corrosion Resistance and Self-Cleaning Property. *J. Mater. Eng. Perform.* **2023**, *32*, 6028–6041.
19. Wang, P.; Yang, Y.; Wang, H.B.; Wang, H.Q. Fabrication of super-robust and nonfluorinated superhydrophobic coating based on diatomaceous earth. *Surf. Coat. Technol.* **2019**, *362*, 90–96.

20. Wang, G.; Chai, Y.M.; Li, Y.F.; Luo, H.J.; Zhang, B.; Zhu, J.F. Sandstone protection by using nanocomposite coating of silica. *Appl. Surf. Sci.* **2023**, *615*, 156193.
21. Su, H.X.; Zhang, J. Modified epoxy resin with SEBS-g-MAH to fabricate crack-free and robust hydrophobic coatings on the surface of PP/SEBS matrix. *Surf. Interfaces* **2022**, *28*, 101662.
22. Liu, J.; Zhang, X.Z.; Wang, R.Y.; Long, F.; Zhao, P.Y. A mosquito-eye-like superhydrophobic coating with super robustness against abrasion. *Mater. Des.* **2021**, *203*, 109552.
23. Dong, K.; Xue, Z.H.; Dong, Y.W.; Lu, Z.Q.; Guan, Z.; Duan, Y.P.; Wang, H.X. Fabrication and properties of n-SiO_2/m-SiO_2/PU superhydrophobic coatings. *J. Phys. Conf. Ser. IOP Publ.* **2023**, *2510*, 012002. [CrossRef]
24. Li, K.; Wang, Y.S.; Wang, X.; Zhang, X.; Zhang, J.H.; Xie, H.S.; Zhang, A.M. Superhydrophobic magnesium oxychloride cement-based composites with integral stability and recyclability. *Cem. Concr. Compos.* **2021**, *118*, 103973.
25. Cui, L.J.; Chen, X.X.; Wang, J.Q.; Qiang, Y.J.; Chen, D.P.; Xiang, T.F. Facile preparation of monolithic superhydrophobic concrete with excellent anti-corrosion property. *Mater. Lett.* **2023**, *344*, 134441.
26. Xiang, T.F.; Liu, J.; Lv, Z.; Wei, F.F.; Liu, Q.W.; Zhang, Y.L.; Ren, H.W.; Zhou, S.C.; Chen, D.P. The effect of silicon-based waterproof agent on the wettability of superhydrophobic concrete and enhanced corrosion resistance. *Constr. Build. Mater.* **2021**, *313*, 125482.
27. Xu, S.S.; Wang, Q.; Wang, N.; Lei, Q.; Song, Q.N. Study of corrosion property and mechanical strength of eco-friendly fabricated superhydrophobic concrete. *J. Clean. Prod.* **2021**, *323*, 129267.
28. Wang, N.; Wang, Q.; Xu, S.S.; Lei, L.L. Green fabrication of mechanically stable superhydrophobic concrete with anti-corrosion property. *J. Clean. Prod.* **2021**, *312*, 127836.
29. Wang, F.J.; Lei, S.; Ou, J.F.; Wen, L. Effect of PDMS on the waterproofing performance and corrosion resistance of cement cement-based materials. *Appl. Surf. Sci.* **2020**, *507*, 145016.
30. Dong, B.B.; Wang, F.H.; Abadikhah, H.; Hao, L.Y.; Xu, X.; Khan, S.A.; Wang, G.; Agathopoulos, S. Simple fabrication of concrete with remarkable self-cleaning ability, robust superhydrophobicity, tailored porosity, and highly thermal and sound insulation. *ACS Appl. Mater. Interfaces* **2019**, *11*, 42801–42807.
31. Zhu, C.X.; Lv, J.; Chen, L.D.; Lin, W.Q.; Zhang, J.; Yang, J.; Yang, J.T.; Feng, J. Dark, heat-reflective, anti-ice rain and superhydrophobic cement concrete surfaces. *Constr. Build. Mater.* **2019**, *220*, 21–28.
32. Cassie, A.B.D.; Baxter, S. Wettability of porous surfaces. *Trans. Faraday Soc.* **1944**, *40*, 546–551.
33. He, Z.K.; Ma, M.; Xu, X.C.; Wang, J.Y.; Chen, F.; Deng, H.; Wang, K.; Zhang, Q.; Fu, Q. Fabrication of superhydrophobic coating via a facile and versatile method based on nanoparticle aggregates. *Appl. Surf. Sci.* **2012**, *258*, 2544–2550.
34. Cui, G.; Zhou, H.X.; Li, L. Research and application of hydrophobic chemical pore bolts on concrete properties. *New Build. Mater.* **2017**, *44*, 28–31.
35. Mazloom, M.; Ramezanianpour, A.A.; Brooks, J.J. Effect of silica fume on mechanical properties of high-strength concrete. *Cem. Concr. Compos.* **2004**, *26*, 347–357.
36. Fallah-Valukolaee, S.; Mousavi, R.; Arjomandi, A.; Nematzadeh, M.; Kazemi, M. A comparative study of mechanical properties and life cycle assessment of high-strength concrete containing silica fume and nanosilica as a partial cement replacement. *Structures* **2022**, *46*, 838–851.
37. Wang, J.B.; Liu, M.L.; Wang, Y.G.; Zhou, Z.H.; Xu, D.Y.; Du, P.; Cheng, X. Synergistic effects of nano-silica and fly ash on properties of cement-based composites. *Constr. Build. Mater.* **2020**, *262*, 120737.
38. Pedro, D.; Brito, J.D.; Evangelista, L. Durability performance of high-performance concrete made with recycled aggregates, fly ash and densified silica fume. *Cem. Concr. Compos.* **2018**, *93*, 63–74.
39. Nie, S.J.; Qi, G.G.; Hu, W.W.; Li, T.; Zhang, H.H.; Wei, K.; Tan, X.; Tu, Y.T.; Li, X.Y.; Guo, R.H.; et al. Study of salt resistance and mechanical strength of robust and non-fluorinated superhydrophobic cement. *Mater. Lett.* **2023**, *339*, 134017. [CrossRef]
40. Momen, G.; Farzaneh, M. Facile approach in the development of icephobic hierarchically textured coatings as corrosion barrier. *Appl. Surf. Sci.* **2014**, *299*, 41–46.
41. He, Q.; Ma, Y.W.; Wang, X.S.; Jia, Y.Y.; Li, K.S.; Li, A.L. Superhydrophobic Flexible Silicone Rubber with Stable Performance, Anti-Icing, and Multilevel Rough Structure. *ACS Appl. Polym. Mater.* **2023**, *5*, 4729–4737.
42. Wang, S.Z.; Baxter, L.; Fonseca, F. Biomass fly ash in concrete: SEM, EDX and ESEM analysis. *Fuel* **2008**, *87*, 372–379. [CrossRef]
43. Jin, W.Z.; Jiang, L.H.; Han, L.; Huang, H.M.; Zhi, F.F.; Yang, G.H.; Niu, Y.L.; Chen, L.; Wang, L.; Chen, Z.Y. Influence of curing temperature on freeze-thaw resistance of limestone powder hydraulic concrete. *Case Stud. Constr. Mater.* **2022**, *17*, e01322.
44. Li, X.D.; Wang, Q.; Lei, L.L.; Shi, Z.Q.; Zhang, M.Y. Amphiphobic concrete with good oil stain resistance and anti-corrosion properties used in marine environment. *Constr. Build. Mater.* **2021**, *299*, 123945.
45. Li, K.; Wang, Y.S.; Zhang, X.; Wu, J.X.; Wang, X.; Zhang, A.M. Multifunctional magnesium oxychloride based composite with stable superhydrophobicity, self-luminescence and reusability. *Constr. Build. Mater.* **2021**, *286*, 122978.
46. She, W.; Zheng, Z.H.; Zhang, Q.C.; Zou, W.Q.; Yang, J.X.; Zhang, Y.S.; Zheng, L.; Hong, J.X.; Miao, C.W. Predesigning matrix-directed super-hydrophobization and hierarchical strengthening of cement foam. *Cem. Concr. Res.* **2020**, *131*, 106029.
47. Li, C.Z.; Jiang, L.H.; Li, S.S. Effect of limestone powder addition on threshold chloride concentration for steel corrosion in reinforced concrete. *Cem. Concr. Res.* **2020**, *131*, 106018.

48. Chen, H.; Feng, P.; Du, Y.; Jiang, J.Y.; Wei, S. The effect of superhydrophobic nano-silica particles on the transport and mechanical properties of hardened cement. *Constr. Build. Mater.* **2018**, *182*, 620–628.
49. Kong, X.M.; Liu, H.; Lu, Z.B.; Wang, D.M. The influence of silanes on hydration and strength development of cementitious systems. *Cem. Concr. Res.* **2015**, *67*, 168–178.

Disclaimer/Publisher's Note: The statements, opinions and data contained in all publications are solely those of the individual author(s) and contributor(s) and not of MDPI and/or the editor(s). MDPI and/or the editor(s) disclaim responsibility for any injury to people or property resulting from any ideas, methods, instructions or products referred to in the content.

Article

Silica Nanoparticle-Infused Omniphobic Polyurethane Foam with Bacterial Anti-Adhesion and Antifouling Properties for Hygiene Purposes

Dongik Cho and Jun Kyun Oh *

Department of Polymer Science and Engineering, Dankook University, 152 Jukjeon-ro, Suji-gu, Yongin-si 16890, Gyeonggi-do, Republic of Korea; dcdk0227@dankook.ac.kr
* Correspondence: junkyunoh@dankook.ac.kr

Abstract: In this study, a method for preventing cross-infection through the surface coating treatment of polyurethane (PU) foam using functionalized silica nanoparticles was developed. Experimental results confirmed that the fabricated PU foam exhibited omniphobic characteristics, demonstrating strong resistance to both polar and nonpolar contaminants. Additionally, quantitative analysis using the pour plate method and direct counting with a scanning electron microscope determined that the treated material exhibited anti-adhesion properties against bacteria. The fabricated PU foam also demonstrated a high level of resistance to the absorption of liquids commonly found in medical facilities, including blood, 0.9% sodium chloride solution, and 50% glycerol. Mechanical durability and stability were verified through repeated compression tests and chemical leaching tests, respectively. The proposed coated PU foam is highly effective at preventing fouling from polar and nonpolar fluids as well as bacteria, making it well-suited for use in a range of fields requiring strict hygiene standards, including the medical, food, and environmental industries.

Keywords: omniphobic; polyurethane foam; bacteria; antifouling; hygiene

Citation: Cho, D.; Oh, J.K. Silica Nanoparticle-Infused Omniphobic Polyurethane Foam with Bacterial Anti-Adhesion and Antifouling Properties for Hygiene Purposes. *Nanomaterials* **2023**, *13*, 2035. https://doi.org/10.3390/nano13142035

Academic Editors: Jose L. Luque-Garcia, Maria Vittoria Diamanti, Massimiliano D'Arienzo, Carlo Antonini and Michele Ferrari

Received: 9 June 2023
Revised: 5 July 2023
Accepted: 7 July 2023
Published: 9 July 2023

Copyright: © 2023 by the authors. Licensee MDPI, Basel, Switzerland. This article is an open access article distributed under the terms and conditions of the Creative Commons Attribution (CC BY) license (https://creativecommons.org/licenses/by/4.0/).

1. Introduction

Recently, with the emergence of the highly contagious novel virus infection, concerns have been raised about infectious diseases around the world. Cross-contamination, in which pathogenic bacteria are transferred through contact with an infected person or material, is a major cause of the spread of such infectious diseases. Medical facilities try to prevent cross-contamination through the use of disposable items as well as strict disinfection and cleaning management. For example, the hospital disposables market is growing, with an average annual growth rate of 9.4% expected until 2026. However, along with the increased use of disposables, the waste generated is also increasing. Medical facilities in the United States generate 4 million tons of waste each year [1]. The cost of disposing this waste accounts for about 20% of the cost of hospital environmental services; thus, not only environmental pollution caused by the waste but also economic costs are becoming a problem [2]. In addition to the use of disposable items, cross-contamination is prevented by methods such as disinfection and cleaning, but for hospital mattresses, bacterial contamination is a problem, even after disinfection. According to French et al., the mattresses used by patients infected with methicillin-resistant *Staphylococcus aureus* (MRSA) remained contaminated with MRSA even after cleaning, and 43% of uninfected patients subsequently placed in these beds became infected with MRSA [3]. Furthermore, based on research conducted by Sharma et al., sulfadiazine antibiotics, which are used as therapeutic agents to inhibit bacterial growth, have certain limitations in their accurate detection methods. Moreover, they not only have the potential to impact the ecosystem but also pose a risk to the human body [4]. These cases suggest that disinfection and cleaning performed as a method for preventing cross-contamination cannot be a fundamental solution.

As patients spend the most time in a medical facility in the hospital bed, the materials most likely to cause infection or the transfer of pathogenic bacteria include mattresses, pillows, and filling materials on which patients lie [5,6]. Polyurethane (PU) foams, which are one of the many types of PU, are mainly used for these materials. PU can be used as a foam, coating, elastomer, or adhesive, depending on the ratio of polyol and isocyanate, owing to its excellent processability and easily controlled physical properties. Among the many forms of PU, PU foams are the most commonly used owing to their high surface area-to-volume ratio, excellent thermal properties, and elasticity [7]. PU foams can be divided into open-cell foams and closed-cell foams, depending on the internal structure [8]. A higher open-cell ratio affects softness and volume shrinkage, and most mattresses (almost 100%) have an open-cell structure that makes them easy to clean [9]. The PU foams of hospital mattresses with a very high percentage of an open-cell structure have a porous network structure, and because of this, they have a very high absorbency for various liquids such as water and oil compared to PU foam with a closed-cell structure. Therefore, for mattresses used by a large number of patients, oil or body fluids generated from the human body can easily penetrate into the cells of the PU foam. This creates an environment conducive to the propagation of biological contaminants such as bacteria, increasing the risk of healthcare-associated infections due to cross-contamination and bacterial transmission [10].

To prevent cross-contamination and bacterial transfer by PU foam used as a mattress material, studies on surface antimicrobial coatings have been conducted to analyze the antibacterial properties of PU foam. A number of studies used antimicrobial agents [11]. For example, Dagostin et al. reported the antimicrobial effect of PU foam treated by adding 0.5 wt% of zinc pyrithione, an antimicrobial agent [12]. Ashjari et al. prepared an antimicrobial mattress in which the growth of bacteria was inhibited by adding CuO to PU foam prepared based on starch [13]. Furthermore, Demirci et al. prepared an open-cell antimicrobial PU foam using quaternary ammonium compounds that had an antimicrobial effect with cationic biocides [14]. Additionally, according to Sienkiewicz et al., bio-based compounds are utilized to provide effective antibacterial and antifungal effects in construction, which is another field where polyurethane foam finds widespread use [15].

The PU foams containing the reported antimicrobial agents can prevent cross-infection by killing bacteria and preventing growth, but the antimicrobial effect decreases over time because of the exhaustion of the antimicrobial agent. The resistance of bacteria to antimicrobial agents may also increase, and when bacterial corpses due to antimicrobial agents form a biofilm on the surface, the antibacterial effect is rapidly reduced, which is inefficient in the long term. A method of preventing cross-infection and bacterial transfer that overcomes these disadvantages is to prepare an anti-adhesion or antifouling surface that prevents bacterial adhesion [16]. The anti-adhesion surface does not kill or inhibit the growth of bacteria but reduces the surface energy of the material to prevent the adhesion of bacteria, thereby reducing the possibility of antimicrobial resistance or biofilm formation [17]. Therefore, to prevent the adhesion of polar contaminants such as bacteria and fungi, as well as nonpolar contaminants (oil and lipid components), an omniphobic surface with repellency to both polar and nonpolar substances is essential.

In contrast to superhydrophobic surfaces with a limited antifouling effect that can only prevent the adhesion of bacteria and polar contaminants, an omniphobic material surface has the advantage of having antifouling effects against both polar and nonpolar contaminants [18]. Omniphobic properties can be determined based on the contact angle between the liquid and the surface; if the contact angles of the surface with water and oil are both greater than 150°, it is considered an omniphobic surface. A lot of research on omniphobic surfaces has been conducted. For example, Pendurthi et al. formed a surface nanomorphology with a re-entrant structure by using a CO_2 laser engraver on the surfaces of stainless steel, aluminum, and glass to create a surface that is repellent to water and oil [19]. Li et al. prepared an omniphobic membrane that can be used as a distiller filter by coating a fluorocarbon surfactant, a fluorinated alkyl silane, and silica (SiO_2) nanoparticles on a polyvinylidene fluoride (PVDF) membrane using a spray coating method [20]. Zhang et al. created omniphobic surfaces by slippery liquid-infused porous surfaces (SLIPSs) with liquid containing ZnO, Co_3O_4, and SiO_2 [21]. Wu et al. fabricated a nanofibrous membrane on a poly(vinylidene fluoride)-*co*-hexafluoropropylene (PVDF-HFP) membrane using the electrospinning method and then fluorinated the surface with a vapor deposition method to impart omniphobic properties [22]. It is difficult to apply this method to products with a short replacement period, such as mattresses, because the control of accurate surface topography is complicated and expensive [23]. In addition, because there is no mention of an experiment confirming bacterial anti-adhesion, a follow-up study on this is necessary. Research confirming the antibacterial properties of PU foam with an omniphobic surface related to the medical field is also required.

In this study, inexpensive reagents and dip-coating methods that could be scalable were applied, with the consideration of inexpensive reagents and a mass production capability for the manufacture of PU foam with omniphobic properties. The aim was an antifouling effect against bacteria as well as polar and nonpolar pollutants that commonly occur in medical environments. To confirm the surface properties of PU foam coated with fluorinated silane-coated silica nanoparticles (FSCSNs), Fourier transform infrared (FTIR) spectroscopy, contact angle measurement, and scanning electron microscopy (SEM) were used. Experiments were conducted to confirm the antibacterial properties of PU foam coated with FSCSNs, in which the Gram-negative bacteria *E. coli* O157:H7 and Gram-positive bacteria *S. epidermidis* were inoculated for the analysis. To confirm the anti-absorption properties of PU foam coated with FSCSNs for polar and nonpolar contaminants, polar liquids (sterile deionized (DI) water, 0.9% sodium chloride solution, and blood) and nonpolar liquids (hexadecane, 50% glycerol solution, and squalene) commonly found in medical facilities were used. A repeated compression test was performed using a counterweight to confirm the durability of the PU foam coated with FSCSNs. Finally, a chemical leaching test was conducted to evaluate the chemical stability. Considering that PU foam is a material that is in contact with the patient's body for the longest time in medical facilities and is highly likely to be exposed to vectors of infectious diseases, the fabricated PU foam is expected to be used as a hospital material that prevents healthcare-associated infections due to cross-infection and bacterial transfer through omniphobic properties achieved using a simple dip-coating method.

2. Materials and Methods

2.1. Preparation of PU Foam Samples

The polyurethane (PU) foam (density of 30.3 kg/m^2, tensile strength of 1.25 kg/m^2, elongation of 130%) used in the experiment was purchased from Kumkang Urethan (Namyangju-si, Republic of Korea). To compare the anti-adhesion effect, untreated PU foam cubes measuring 2 cm × 2 cm × 2 cm were cut, soaked in ethanol (95%, Daejung Chemicals, Siheung-si, Republic of Korea) for 5 min, and dried in an oven at 60 °C for 20 min.

2.2. Fluorinated Silica Nanoparticles Suspension

First, 300 mg of silica (SiO_2; Sigma-Aldrich Co., St. Louis, MO, USA) nanoparticles with an average diameter of ca. 20 nm were added to 50 mL hexane (Daejung Chemicals, Siheung-si, Republic of Korea) and suspended for 20 min with a probe-type ultrasonicator. Then, 5 mM of trichloro(1H,1H,2H,2H-heptadecafluorodecyl)silane (HFTCS, Tokyo Chemical Industry, Tokyo, Japan) was added to the suspension and suspended for another 20 min to obtain FSCSNs. The FSCSNs suspension was left at room temperature for 2 h to sufficiently react the silica nanoparticles with the silane.

2.3. Dip-Coating

The dip-coating method was used to coat FSCSNs on PU foam. PU foam cut into 2 cm × 2 cm × 2 cm cubes was dipped in the FSCSNs suspension. Dipping in the suspension for 30 s followed by drying at room temperature for 30 s was repeated five times. The coating was performed at a speed of 0.5 cm/s so that the FSCSNs were uniformly coated. Finally, the PU foam coated with FSCSNs was dried at room temperature for 24 h to obtain FSCSNs-coated PU foam.

2.4. Surface Characterization

To confirm the chemical interaction and trifluoromethyl ($-CF_3$) functional groups between HFTCS functionalized silica nanoparticles inside and outside the coated PU foam, we used FTIR spectroscopy. FTIR spectra were measured using a spectrometer (Nicolet iS10, Thermo Fisher Scientific, Waltham, MA, USA) and analyzed using OMNIC software (Thermo Fisher Scientific, Waltham, MA, USA).

To measure the wetting characteristics, the static contact angle of untreated (bare) PU foam and coated PU foam was measured using the sessile drop technique. The experiment involved dropping droplets of water and hexadecane of the same volume (5 µL) onto the foam surface five times at room temperature and then determining the average value from the measurements. Contact angles were analyzed using ImageJ software (National Institutes of Health, Bethesda, MD, USA) via the low-bond axisymmetric drop shape analysis (LBADSA) plugin [24].

Finally, SEM (S-4700s; Hitachi, Tokyo, Japan) was used to check the surface morphology of the coated PU foam. To enable the conductivity of the sample during the SEM measurement, a layer of platinum with a thickness of 20 nm was applied prior to the SEM analysis. The SEM was operated with a voltage of 20 kV and a current of 10 µA.

2.5. Bacterial Cultures

In this study, two types of bacteria were used: Gram-negative *Escherichia coli* O157:H7 (ATCC 25922) and Gram-positive *Staphylococcus epidermidis* (ATCC 12228). *E. coli* O157:H7 and *S. epidermidis* were transferred from tryptic soy agar slants (TSA; Becton, Dickinson and Co., Franklin Lakes, NJ, USA) to a culture tube containing 9.0 mL of tryptic soy broth (TSB; Becton, Dickinson and Co., Franklin Lakes, NJ, USA) using a loopful (10 µL) and cultured. The tubes of all strains were incubated aerobically at 37 °C for 24 h without shaking. A second transfer was conducted by transferring a loopful culture to a fresh TSB culture medium and then incubating under the same conditions. The experiment utilized final concentrations of the two bacteria ranging from 8.8 to 9.2 log CFU/mL.

2.6. Bacterial Adhesion Assay

To evaluate whether the anti-adhesion effect of the foam persists even after prolonged exposure to bacteria, bare and FSCSNs-coated PU foam samples were immersed in 9 mL of a bacterial suspension at room temperature for 1 h and 8 h. After immersion, the samples were removed from the bacterial suspension quickly and then transferred to a conical tube containing sterile DI water (9 mL) for a further bacterial adhesion assay. The bacterial adhesion assay was conducted in a suitable biological safety cabinet with sterile conditions.

The adhesion of bacteria on both the bare and FSCSNs-coated PU foam surfaces was evaluated using the pour plating method for plate counting, as well as direct counting on the PU foam surfaces using SEM. For plate counts, PU foam samples that were inoculated using conical tubes for 1 h and 8 h were each vortexed in sterile DI water for 1 min at 3000 RPM to detach bacteria from the PU foam surfaces. After that, the DI water (1 mL) containing bacteria detached from the sample was added to 0.1% (w/v) peptone water (9 mL), serial dilution was performed, and the final solution was spread on a TSA plate. The bacterial density was counted after 24 h of aerobic incubation at 37 °C and refers to the density of bacteria adhering to the PU foam surfaces. All experiments were replicated three times.

Counting the number of bacteria attached to the surface of the PU foam using SEM was performed using PU foam samples immersed in the inoculum for 1 h and 8 h. Before imaging by SEM, bacteria were inactivated using ethanol, and 20 nm of platinum coating was applied to ensure electrical conductivity for the SEM measurement. For statistical reliability, more than five areas of 100 μm × 100 μm were observed in three samples of the same type of PU foam. SEM micrographs were analyzed using ImageJ to quantify the attachment of *E. coli* O157:H7 and *S. epidermidis* to PU foam surfaces.

2.7. Absorption Capacity Test

The test method for measuring absorption capacity was based on ASTM F726-99 (i.e., the standard test method for the sorbent performance of adsorbents). First, in the nonpolar liquid absorption comparison test, 1 mL of hexadecane was added to a Petri dish, and the weighed bare PU foam and FSCSNs-coated PU foam were each immersed in hexadecane for 1 min. Subsequently, the PU foam was removed and left to dry at room temperature for 5 s. The saturated PU foam was then placed on a pre-weighed Petri dish, and the weight of the saturated PU foam was recorded by subtracting the weight of the Petri dish. The test was conducted in the same way using 50% glycerol and squalene, which are other nonpolar liquids used in hospitals. The absorption comparison tests for polar liquids (DI water, 0.9% sodium chloride solution, and blood), which are often used in hospitals, were also conducted using a similar method. The absorption capacity of the PU foam was calculated according to Equation (1), where M_d is the initial dry weight of the PU foam, and M_w is the weight of the saturated PU foam after absorbing the liquid. All absorption tests were performed under the same conditions at room temperature, and the average absorption capacity was calculated by repeating three times. In addition, if the result deviated from the average by 10%, a retest was conducted using a new sample.

$$\text{Absorption capacity (g/g)} = \frac{(M_w - M_d)}{M_d} \quad (1)$$

2.8. Compression Test

A compression test was performed on FSCSNs-coated PU foam to confirm that the omniphobic properties were maintained even under repeated stress due to the characteristics of mattresses, which must withstand the weight of the human body for a long period. The mechanical properties were measured by repeatedly applying the same pressure in the vertical direction with a 500 g weight connected by string to the FSCSNs-coated PU foam with a size of 1.5 cm × 1.5 cm × 1.5 cm. The coated PU foam was subjected to a total of 100 compressions, while assessing its ability to repel liquids. To confirm the omniphobic properties of the foam, droplets of water and hexadecane (5 μL each) were applied five times each after every ten compressions, and the resulting contact angles were measured and averaged.

2.9. Chemical Leaching Test

The chemical stability of the FSCSNs-coated foam was determined by checking whether there was chemical leaching on the surface of the foam over time when the FSCSNs-coated foam was immersed in a polar or nonpolar liquid. Polar and nonpolar liquids were tested using 0.9% sodium chloride solution and 50% glycerol, liquids commonly used in hospitals. FTIR spectroscopy, with a detection limit of <1 ppm, was employed as the analysis method. The immersion condition was set to a maximum of 14 d, and the results were compared by performing analysis every 7 d.

2.10. Statistical Analysis

The bacterial adhesion assay results were analyzed after log-transformation. The significance level was set to $p < 0.05$ in the results of the bacterial attachment analysis. For this purpose, two-way analysis of variance with Tukey's post hoc test was used. Statistical analyses were performed using the Analysis ToolPak in Excel (Microsoft Corp., Redmond, WA, USA) via statistical packages.

3. Results and Discussion
3.1. Characterization of FSCSNs-Coated PU Foam Surfaces

Omniphobic PU foam with both superhydrophobic and superoleophobic properties was manufactured by coating the surface of the PU foam with FSCSNs by silanization through a condensation reaction between −OH and −Cl groups on the surface of the PU foam (Figure 1). Bacterial adhesion can be affected when nano or microscopic roughness is applied to the surface through chemical or physical methods. If the surface roughness unit is larger than the size of the bacteria, the bacteria penetrate the topography, such as valleys, edges, and pits between the surface roughness, thereby increasing the contact area with enhanced bacterial adhesion on the surface as well as bacterial colonization. In addition, PU foam has a three-dimensional porous structure and a large surface area, allowing air to enter and exit smoothly and creating a suitable environment for bacteria to inhabit. Therefore, in this study, FSCSNs were coated on the porous surface of PU foam to lower the surface energy and to impart nanoscale surface roughness to achieve a high repellency against contaminants. These properties are due to HFTCS, a fluorinated silane used for surface modification, and, more specifically, due to the bond between carbon and fluorine in the chain structure of HFTCS. A carbon–fluorine (C−F) bond is a bond between carbon with an electronegativity of 2.5 and fluorine with an electronegativity of 4.0, which is a very strong bond with a large difference in electronegativity. That is, the electron density around the carbon atom is reduced and that around the fluorine atom is increased to create positive and negative charges, respectively; the charges cancel each other out so that the net charge of the perfluorinated molecule is zero. The high electronegativity of fluorine reduces the polarizability of the bond, limiting its susceptibility to van der Waals interactions (related to the magnetic susceptibility, which is the ratio of the strength of magnetization of a material to the strength of a magnetic field), leading to very low surface energies of perfluorinated materials [25]. FTIR analysis was performed to confirm the presence of trifluoromethyl functional groups on the inside and outside of the coated PU foam (Figure 2a). Although no peak was observed in the corresponding wavelength band in the bare PU foam, in the coated PU foam, a peak was observed at a wavelength of 1050 cm^{-1} for samples both inside and outside the treated PU foam, confirming that this corresponded to the C−F stretching peak. The presence of a peak at approximately 1050 cm^{-1} confirms the occurrence of C−F stretching. Furthermore, the observed peak shift of approximately 8 cm^{-1} towards lower wavelengths can be attributed to the reduction in C−F bond strength resulting from the functionalization process involving silica nanoparticles [26]. Additionally, we observed weak vibrations at 950 cm^{-1} both inside and outside the regions of the PU foam coated with FSCSNs. These vibrations were verified to be the result of an asymmetric vibration caused by Si−OH groups introduced by the presence of SNPs [27]. Therefore, it was confirmed that the porous surface of the PU foam was coated with FSCSNs [28].

The surface morphology modified by FSCSNs generated on the porous surface of PU foam and the chemical modification by HFTCS create a synergistic effect for repelling bacteria [29]. FSCSNs form a hierarchical structure on the PU foam surface. When an air layer is formed between uniformly aggregated FSCSNs and bacteria adhere to the surface, the air pocket formed between the surface and the bacteria causes an air-pocket effect, which causes the bacteria to come into contact with the surface in a nonlinear manner, eventually affecting the growth of the bacteria [30]. In this study, to compare the wettability of bare PU foam and PU foam coated with FSCSNs to that of water and oil, the static contact angle was measured for DI water and hexadecane (Figure 2b). The water contact angle of the bare PU foam was 123.2°, indicating hydrophobicity, but the water was observed to be gradually absorbed into the PU foam over time. Because the hexadecane droplet was absorbed into the PU foam as soon as it fell, the contact angle for hexadecane was indicated as <10°. In contrast, the contact angles of the PU foam coated with FSCSNs were 153.5° and 153.4° for water and hexadecane, respectively. As these contact angles indicate super water repellency and super oil repellency, respectively, the PU foam coated with FSCSNs had omniphobic characteristics due to the effect of low surface energy and nanoscale surface roughness generated by FSCSNs. The contact angle of a surface can be influenced by various physical factors, including the roughness, texture, and hierarchical structure. Simultaneously, chemical factors such as surface heterogeneity can also have a substantial impact on determining the contact angle [31,32].

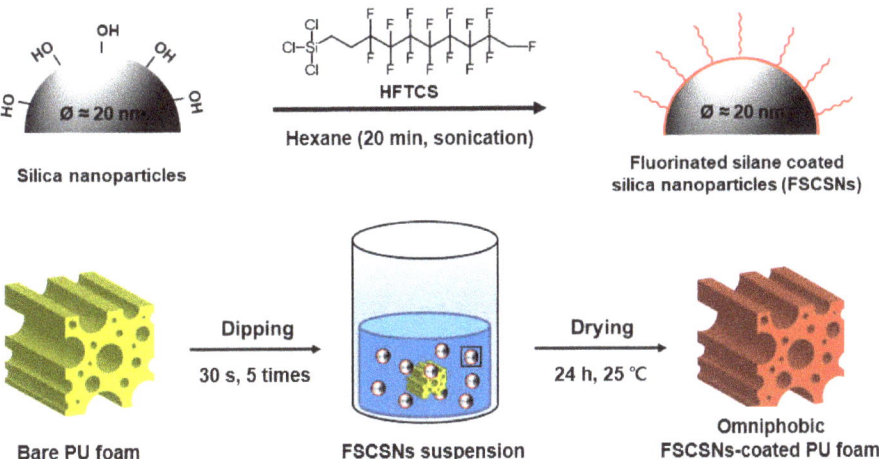

Figure 1. Schematic illustration of the synthesis of FSCSNs through the silanization of HFTCS and silica nanoparticles. Fabrication of FSCSNs-coated PU foam with omniphobic properties using FSCSNs suspension and the dip-coating method.

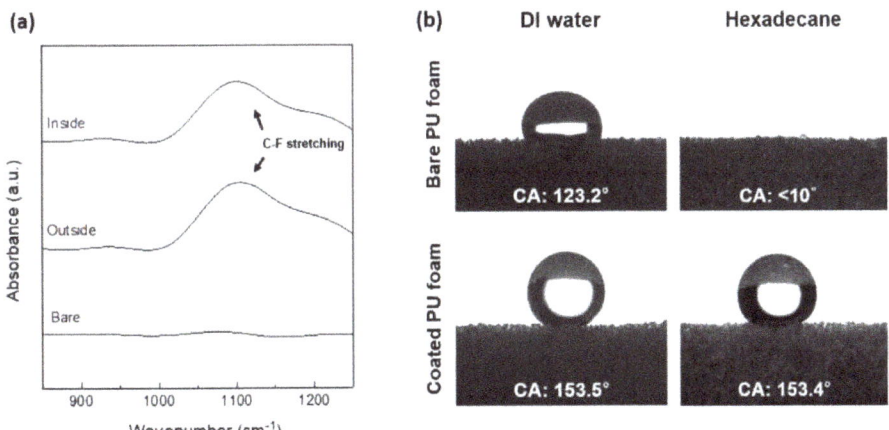

Figure 2. (a) C−F stretching region from the FTIR spectra of the inside and outside regions of the FSCSNs-coated PU foam and the bare PU foam. (b) Static contact angle (CA) measurement results for bare PU foam and FSCSNs-coated PU foam.

Furthermore, an experiment was conducted to confirm whether both the inside and outside of the coated PU foam had omniphobic properties. Figure 3a shows photographs taken after dropping DI water (blue) and hexadecane (red) on the bare and coated PU foam samples to examine the liquid repellency of the inside and outside of the PU foam. As a result, the liquid dropped on the bare PU foam permeated inside, whereas the liquid dropped on the FSCSNs-coated PU foam maintained its spherical shape. Figure 3b shows the results of measuring the static contact angles of the inside and outside of the coated PU foam. The results confirmed that the contact angle of both the inside and outside of the coated PU foam was 150° or more, indicating that it had omniphobic properties.

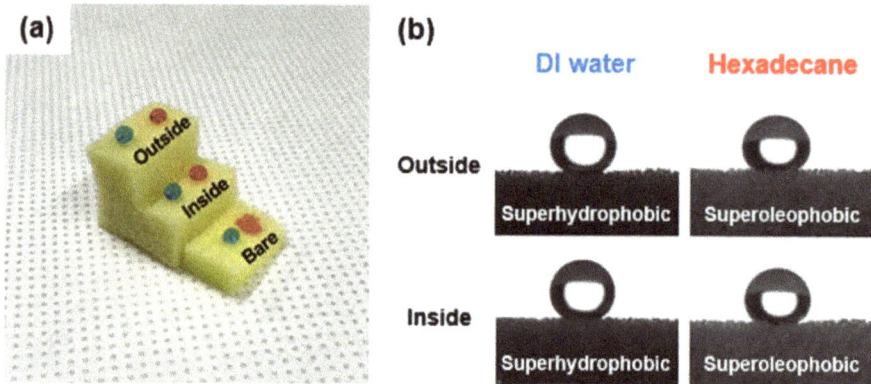

Figure 3. (a) Photographs of polar and nonpolar liquid (DI water: blue, hexadecane: red) on the inside and outside regions of FSCSNs-coated foam and bare foam. (b) Photographs of the static contact angle for polar and nonpolar liquids inside and outside the FSCSNs-coated foam.

The surface morphology of PU foam with a porous structure was observed using SEM (Figure 4). On the surface of the bare PU foam, the entire surface structure had an open-cell structure, and the surface of the internal network structure was smooth. In contrast, as shown in the enlarged photograph of the surface of the PU foam coated with FSCSNs, a hierarchical structure was formed on the surface of the PU foam due to the FSCSNs. However, despite these coatings, it was confirmed that the surface morphology of the PU foam coated with FSCSNs also maintained an open-cell structure. The coating did not affect the characteristics of the mattress, as the softness, volumetric shrinkage, and shock mitigation of the mattress were determined by the open-cell structure [33].

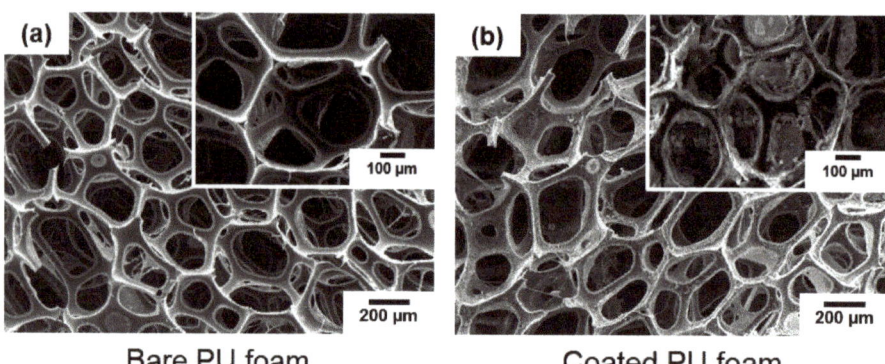

Figure 4. SEM micrographs of (a) bare and (b) FSCSNs-coated PU foams showing open-cell structures.

3.2. Bacterial Attachment to FSCSNs-Coated PU Foam Surfaces Characterized by Plating Counting

The bacterial adhesion behavior on the bare PU foam and the PU foam coated with FSCSNs over time was confirmed using the pour plate method. The plate counting results of the PU foam samples inoculated with (exposed to) the bacterial suspension for 1 h and

8 h are quantitatively represented graphically in Figure 5. First, the results of inoculation for 1 h, as shown in the left graph of Figure 5a, indicate that the mean population of bacteria adhered to the PU foam inoculated for 1 h in *E. coli* O157:H7 suspension was 7.3 log CFU/mL for the bare PU foam and 4.5 log CFU/mL for the PU foam coated with FSCSNs. In addition, in the results of inoculation with *S. epidermidis* shown in the right graph, the mean population of bacteria adhered to the PU foam was 7.4 log CFU/mL for the bare PU foam and 5.0 log CFU/mL for the PU foam coated with FSCSNs. In other words, as a result of immersion in the bacterial suspension for 1 h, the bacteria adhering to the coated PU foam surface were reduced by 98.9%. The trend of the 1 h adhesion test results was similar for the 8 h test. As shown in Figure 5b, the mean population of bacteria adhered to the PU foam inoculated for 8 h in *E. coli* O157:H7 suspension was 7.4 log CFU/mL for the bare PU foam and 5.8 log CFU/mL for the coated PU foam. In addition, as a result of inoculation with *S. epidermidis*, the mean population of bacteria adhered to the PU foam was 7.5 log CFU/mL for the bare PU foam and 5.7 log CFU/mL for the PU foam coated with FSCSNs. Similar to the results of inoculation for 1 h, the PU foam coated with FSCSNs showed a 97.4% decrease in the degree of adhesion after 8 h of bacterial exposure compared to the bare PU foam. This confirms the antimicrobial properties of the PU foam coated with FSCSNs; the results of the pour plate method indicated an average of a 1–2 log units decrease compared to the bare PU foam. The PU foam coated with FSCSNs could prevent the proliferation of bacteria even during the proper sleeping time (i.e., 8 h), when the body is in contact with the mattress for a long time.

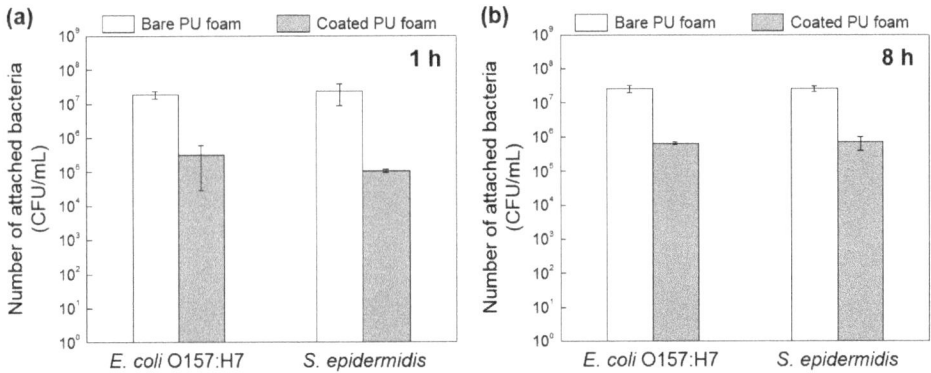

Figure 5. Comparison of bacterial adhesion between bare PU foam and FSCSN-coated PU foam through quantitative results after exposure to bacteria for (**a**) 1 h and (**b**) 8 h.

3.3. Bacterial Attachment to FSCSNs-Coated PU Foam Surfaces Characterized by SEM

In addition to the pour plate method, the number of bacteria adhered per unit area was directly counted using SEM to confirm the antimicrobial properties of the treated PU foam, and the distribution was quantified. After inoculating with the bacteria for 1 h and 8 h, the bacteria adhered to the two types of PU foam surfaces were compared. As shown in Figure 6, a large number of bacteria of both types adhered to the surface of the bare PU foam. In contrast, on the surface of the PU foam coated with FSCSNs, adhered bacteria could not be identified, and only uniform and hierarchical structures generated by FSCSNs could be identified. The number of bacteria adhered to the PU foam samples inoculated with *E. coli* O157:H7 for 1 h was 6.1 log cells/mm^2 for the bare PU foam and 4.5 log cells/mm^2 for the PU foam coated with FSCSNs, showing a 97.4% decrease. The number of *S. epidermidis* bacterial cells adhered to the PU foams was 6.1 log cells/mm^2 for the bare PU foam and 4.3 log cells/mm^2, showing a 98.3% decrease.

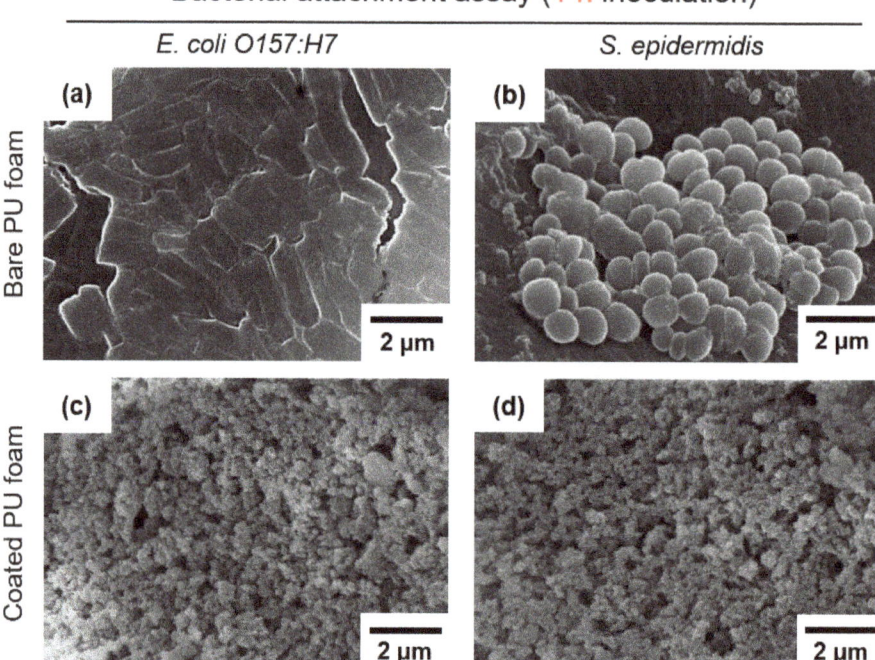

Figure 6. SEM micrographs of (**a**) *E. coli* O157:H7 attached to bare PU foam, (**b**) *S. epidermidis* attached to bare PU foam, (**c**) *E. coli* O157:H7 attached to FSCNSs-coated PU foam, and (**d**) *S. epidermidis* attached to FSCNSs-coated PU foam after 1 h exposure to bacteria.

After immersing the two types of PU foam in the bacterial suspension for 8 h, reflective of the sleeping time, the number of adhered bacteria was also counted. Looking at the results shown in Figure 7, the number of bacteria adhered to the PU foams inoculated with *E. coli* O157:H7 for 8 h was 6.2 log cells/mm^2 for the bare PU foam and 4.7 log cells/mm^2 for the coated PU foam, showing a 96.7% decrease. The number of *S. epidermidis* bacteria adhered to the PU foams was 6.1 log cells/mm^2 for the bare PU foam and 4.5 log cells/mm^2 for the coated PU foam, showing a 97.4% decrease. Therefore, as a result of analyzing the bacterial adhesion properties, the PU foam coated with FSCSNs had excellent anti-adhesion properties, having a significant ($p < 0.5$) reduction of 1–2 log units or more for Gram-negative and Gram-positive bacteria. As a result of indirect counting using the pour plate method described in Section 3.2 and direct counting using SEM, when omniphobic PU foam coated with FSCSNs was exposed to bacterial suspensions containing *E. coli* O157:H7 and *S. epidermidis* for 1 h and 8 h, it exhibited anti-adhesion properties against bacteria, with smaller numbers of bacteria adhered to the surface than for the bare PU foam. This phenomenon can be explained by the wetting transition of the PU foam surface from the Wenzel state to the Cassie–Baxter state caused by the FSCSN coating [34]. The transition of the PU foam surface to the Cassie–Baxter state means that the real contact area is reduced by air pockets generated when hydrophilic bacteria come into contact with the PU foam surface [35]. In other words, the anti-adhesion effect can be explained by hydrophilic and hydrophobic effects [36,37]. The surface layer of *E. coli* O157:H7 and *S. epidermidis* has a hydrophilic property at a water contact angle between 16° and 57° [38,39]. When bacteria with such a hydrophilic cell surface come into contact with a nonpolar surface, intermolecular interactions repel them, thereby hindering bacterial adhesion [40].

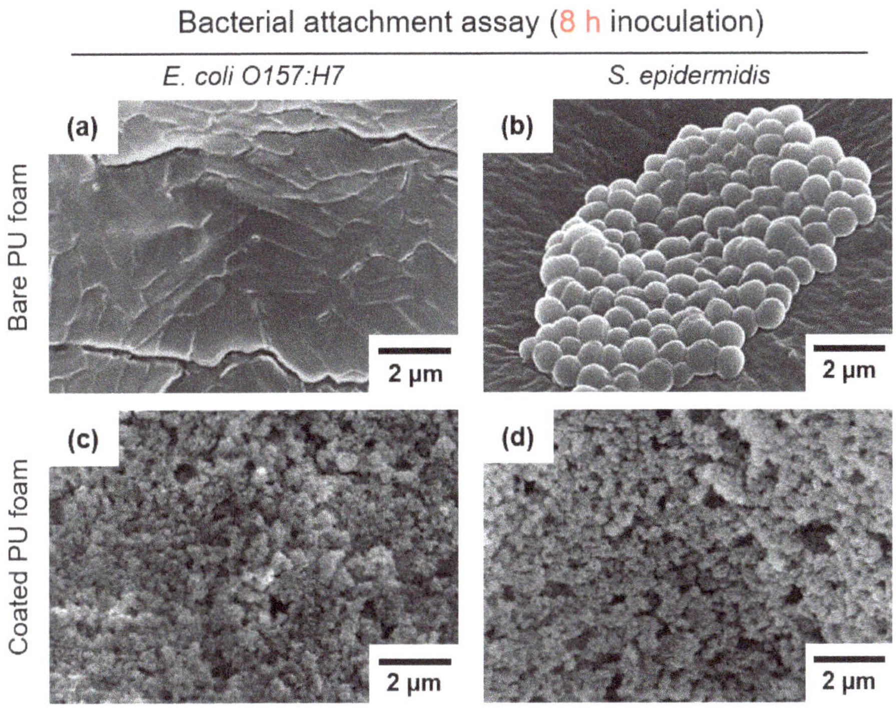

Bacterial attachment assay (8 h inoculation)

Figure 7. SEM micrographs of (**a**) *E. coli* O157:H7 attached to bare PU foam, (**b**) *S. epidermidis* attached to bare PU foam, (**c**) *E. coli* O157:H7 attached to FSCNSs-coated PU foam, and (**d**) *S. epidermidis* attached to FSCNSs-coated PU foam after 8 h exposure to bacteria.

3.4. Anti-Absorption Test of FSCSNs-Coated PU Foam

An absorption test was conducted to confirm the ability of the fabricated PU foam to prevent contamination by polar and nonpolar liquids. The absorption capacities for the liquids were calculated using Equation (1), and the results are shown in Figure 8. As for the polar liquids, the average absorption capacity of DI water was 4.2 (g/g) for the bare PU foam and 0.1 (g/g) for the coated PU foam. The average absorption capacity of sodium chloride solution was 5.8 (g/g) for the bare PU foam and 0.1 (g/g) for the coated PU foam. Finally, the average absorption capacity of blood was the highest at 8.7 (g/g) for the bare PU foam and 0.1 (g/g) for the coated PU foam. From these results, the PU foam coated with FSCSNs showed a very high average reduction rate of the absorption capacity of polar liquids of more than 99%. Absorption capacity tests for nonpolar liquids also showed similar results. The average absorption capacity of hexadecane was 7.1 (g/g) for bare PU foam and 0.1 (g/g) for coated PU foam. The average absorption capacity of glycerol solution was 14.9 (g/g) for bare PU foam and 0.1 (g/g) for coated PU foam. Finally, the average absorption capacity of squalene was the highest at 26.5 (g/g) for the bare PU foam and at 0.1 (g/g) for the coated PU foam. In other words, PU foam coated with FSCSNs also showed an average absorption reduction rate of more than 99% for nonpolar liquids, confirming excellent water and oil absorption prevention capabilities. Therefore, as the elasticity of the PU foam did not decrease due to the absorption of liquid contaminants, it could be used as a mattress material for a long period of time [41,42].

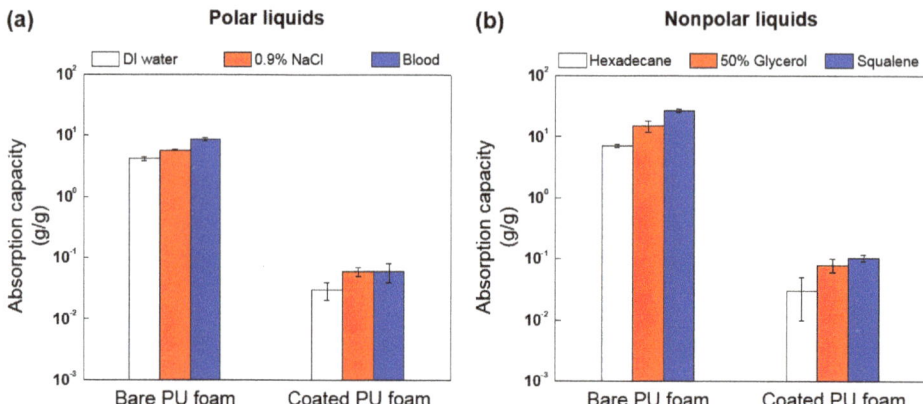

Figure 8. Absorption capacity of bare and FSCSNs-coated PU foams for (**a**) polar liquids (DI water, 0.9% NaCl, and blood) and (**b**) nonpolar liquids (hexadecane, 50% glycerol, and squalene).

The liquids used in the absorption experiment are all commonly found in medical facilities. Body fluid consists of 1% sodium chloride and various salts and organic components, and body sebum contains an average of 6% to 13% squalene, although there are differences according to gender and age [43,44]. Therefore, a mattress that comes in close contact with the body must have properties that prevent the absorption of such body fluids. In addition, glycerol is a low-toxicity substance that is used in the food, medicine, and cosmetic industries and is the most effective and widely used optical clearing agent among various types of alcohol [45,46]. According to a study by Lai et al., medical facilities use a 50% glycerol solution that is safe for skin application in vivo as a light scattering attenuator. Based on the literature, the concentration of glycerol was set to 50% for absorption experiments [47].

3.5. Mechanical Durability of FSCSNs-Coated PU Foam

Coated omniphobic PU foam is a mattress material, and due to its nature, it must maintain the omniphobic property of high water and oil repellency even under frequent pressure for a long period of time. Therefore, in this study, repeated compression tests were conducted to confirm the mechanical durability of the PU foam coated with FSCSNs. The weights and sizes of the samples used in the experiment were selected with reference to previous studies showing that the average interfacial pressure applied to a mattress by an adult male weighing 80 kg is 4 kPa to 8 kPa [48]. Based on this, the size of the PU foam coated with FSCSNs was set to 1.5 cm × 1.5 cm × 1.5 cm, and a pressure similar to that described in the literature was applied by placing a 500 g weight. As shown in Figure 9, to confirm whether the omniphobic properties were maintained after applying pressure for a certain number of times, the contact angles for water and hexadecane were measured every 20 cycles, and the results were summarized in a graph. The contact angles for water and hexadecane showed a high repellency of more than 150° throughout the experiment. The contact angle measured after applying pressure 100 times was 154.6° for DI water and 153.7° for hexadecane, confirming that the omniphobic properties were maintained. The porous ratio and depth present in polyurethane foam do not affect its resilience property [49]. Therefore, an omniphobic PU foam with a high proportion of an open-cell structure maintains its resilience related to mechanical durability. As a result, it retains the performance and omniphobic characteristics even under repeated compression.

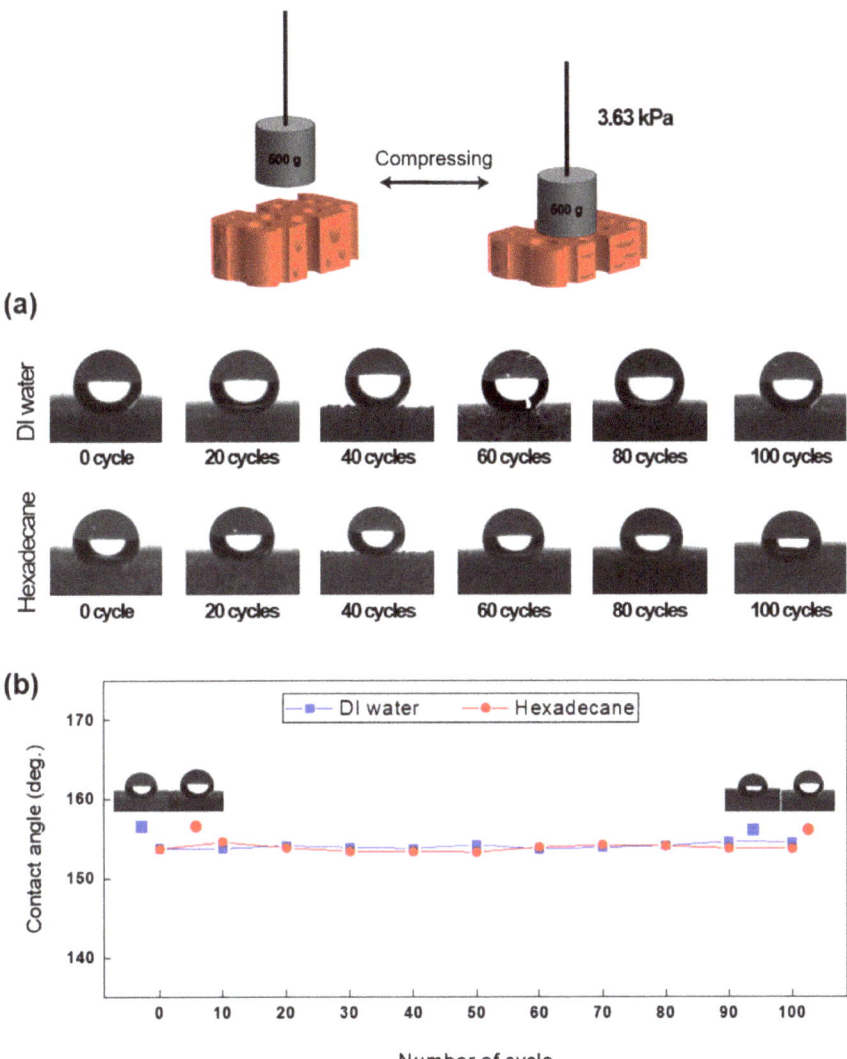

Figure 9. (a) Sequential photos showing the water contact angle and oil contact angle after multiple iterations of a compression test. (b) The contact angle measurements graph of FSCSNs-coated foam during the cyclic compression test.

3.6. Chemical Leaching Test

Because PU foam is used in an environment where there is frequent contact with the body, such as a mattress in a medical facility, the potential toxicity of the PU foam coating must be evaluated. Therefore, chemical leaching should not occur due to the separation or decomposition of fluorine groups on the surface of silica nanoparticles functionalized with HFTCS. Therefore, in this study, leaching experiments were conducted using FTIR for two types of liquids to confirm the chemical stability of the PU foam coated with FSCSNs (Figure 10). The graph in Figure 10a shows the results of leaching after immersing the coated PU foam in a 0.9% sodium chloride solution, a polar liquid often used in medical facilities, for up to 14 d. When compared with the FSCSN suspension as the control group, unbound FSCSN molecules exhibited a C−F stretching peak around 1050 cm^{-1}. However, no peak was observed for at least 14 d in the 0.9% sodium chloride solution in which the coated PU foam was immersed. When a saline solution containing Na and Cl ions is present, each ion causes a specific arrangement of water molecules, impacting the OH stretching band, which is primarily observed around 3300 cm^{-1}. Consequently, no peak is observable for the 0.9% NaCl solution within the measured wavelength range of 850 cm^{-1} to 1250 cm^{-1} [50].

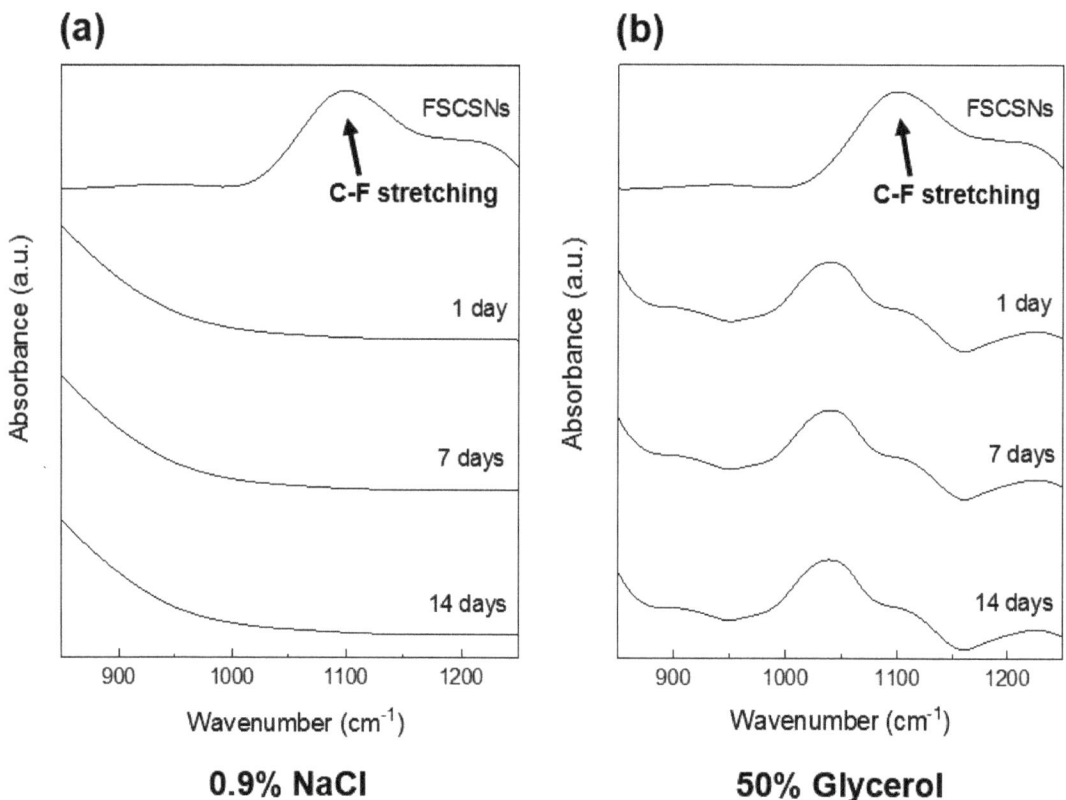

Figure 10. FTIR spectra of aliquots collected from samples of the FSCSNs-coated foam that had been immersed for up to 14 d in (**a**) 0.9% NaCl and (**b**) 50% glycerol, showing the absence of C−F stretching.

The graph in Figure 10b shows the results of an experiment confirming chemical stability using a 50% glycerol solution, which is a nonpolar liquid. Except for the C−O−C stretching peak of the 50% glycerol solution around 1103 cm^{-1}, no peaks were observed, including a C−F stretching peak around 1050 cm^{-1}. This means that no chemical leaching occurred in the 50% glycerol solution in which the PU foam coated with FSCSNs was immersed [51]. Furthermore, glycerol-containing water exhibits primary and secondary hydroxyl groups within the range of 1000–1100 cm^{-1}, leading to observable peaks at those specific wavelengths [52]. Moreover, the C−F stretching shows a strong peak at 1050 cm^{-1}, attributed to the overlap between the Si−O−Si bond and the −CF$_3$ symmetric stretching. However, the influence of NaCl and glycerol on the C−F stretching is minimal, as no significant changes were observed [53]. No chemicals were detected within the detection limit of 1 ppm for at least 14 d in either liquid used, indicating that the fabricated PU foam is chemically stable in both polar and nonpolar liquids.

4. Conclusions

To solve the problem of cross-contamination, which has become more important due to the emergence of infectious diseases, this study aimed to prevent the contamination of mattresses that come in close contact with the human body in medical facilities. The surface of PU foam, the main material used in such mattresses, was coated with silica nanoparticles functionalized with fluorinated silane to reduce surface energy and form nanoscale surface roughness. The key results were the following: (1) The treated PU foam exhibited high repellency to liquids such as water and oil; (2) the adhesion of harmful bacteria such as *E. coli* O157:H7 and *S. epidermidis* was reduced by more than 90% per unit area (1–2 log units) compared to untreated PU foam; and (3) a high antifouling effect against polar and nonpolar liquid pollutants common in medical facilities was observed. The durability of the prepared PU foam was confirmed through repeated compression experiments. The experiment was conducted considering the average interfacial pressure applied to the mattress by adults, and as a result, the omniphobic properties were maintained even after repeated compression up to 100 times. Furthermore, to confirm the chemical stability, leaching experiments were conducted under the conditions of a 0.9% sodium chloride solution and a 50% glycerol solution. Peaks corresponding to carbon and fluorine bonds could not be identified in the two solutions for up to 2 weeks. Based on the findings of this study, the PU foam coated with FSCSNs has a high antifouling effect due to water and oil repellency as well as anti-adhesion properties against bacteria. Moreover, the proposed coating method is simple and scalable and employs inexpensive reagents. In addition to the medical field, it is expected to be applicable to various industries that require hygiene, such as food, electronics, and textiles.

Author Contributions: The manuscript was written with contributions from both authors. J.K.O. was responsible for the experiment design and conceptualization. D.C. conducted the laboratory work and analyzed the data. The manuscript was jointly written by D.C. and J.K.O. and was subsequently reviewed and edited by J.K.O. All authors have read and agreed to the published version of the manuscript.

Funding: This research received no external funding.

Data Availability Statement: The data that support the findings of this study are available upon request from the authors.

Acknowledgments: The present research was conducted by the research fund of Dankook University in 2023.

Conflicts of Interest: The authors declare no conflict of interest.

References

1. Otter, J.A.; Nowakowski, E.; Salkeld, J.A.G.; Duclos, M.; Passaretti, C.L.; Yezli, S.; Ross, T.; Carroll, K.C.; Perl, T.M. Saving costs through the decontamination of the packaging of unused medical supplies using hydrogen peroxide vapor. *Infect. Control Hosp. Epidemiol.* **2013**, *34*, 472–478. [CrossRef]
2. Conrardy, J.; Hillanbrand, M.; Myers, S.; Nussbaum, G.F. Reducing medical waste. *AORN J.* **2010**, *91*, 711–721. [CrossRef]
3. French, G.L.; Otter, J.A.; Shannon, K.P.; Adams, N.M.T.; Watling, D.; Parks, M.J. Tackling contamination of the hospital environment by methicillin-resistant *Staphylococcus aureus* (MRSA): A comparison between conventional terminal cleaning and hydrogen peroxide vapour decontamination. *J. Hosp. Infect.* **2004**, *57*, 31–37. [CrossRef]
4. Sharma, T.S.K.; Jana, J.; Bhamu, K.C.; Song, J.; Sivaselvam, S.; Van Tam, T.; Kang, S.G.; Chung, J.S.; Hur, S.H.; Choi, W.M. Rational synthesis of alkaline earth metal vanadates: Structural origin of MgVO3 honeycomb lattice system and its electrochemical analysis for the detection of sulfadiazine. *Chem. Eng. J.* **2023**, *464*, 142673. [CrossRef]
5. Creamer, E.; Humphreys, H. The contribution of beds to healthcare-associated infection: The importance of adequate decontamination. *J. Hosp. Infect.* **2008**, *69*, 8–23. [CrossRef]
6. Bartley, J.M.; Olmsted, R.N.; Hass, J. Current views of health care design and construction: Practical implications for safer, cleaner environments. *Am. J. Infect. Control* **2010**, *38*, S1–S12. [CrossRef]

7. Seyed Dorraji, M.S.; Rasoulifard, M.H.; Shajeri, M.; Ashjari, H.R.; Azizi, M.; Rastgouy-Houjaghan, M. The role of prepared ZnO nanoparticles on improvement of mechanical and antibacterial properties of flexible polyurethane foams: Experimental modeling. *Polym. Bull.* **2018**, *75*, 1519–1533. [CrossRef]
8. Rodriguez-Perez, M.A.; Álvarez-Láinez, M.; de Saja, J.A. Microstructure and physical properties of open-cell polyolefin foams. *J. Appl. Polym. Sci.* **2009**, *114*, 1176–1186. [CrossRef]
9. Harikrishnan, G.; Patro, T.U.; Khakhar, D.V. Polyurethane foam-clay nanocomposites: Nanoclays as cell openers. *Ind. Eng. Chem. Res.* **2006**, *45*, 7126–7134. [CrossRef]
10. Livshiz-Riven, I.; Borer, A.; Nativ, R.; Eskira, S.; Larson, E. Relationship between shared patient care items and healthcare-associated infections: A systematic review. *Int. J. Nurs. Stud.* **2015**, *52*, 380–392. [CrossRef]
11. Dagostin, V.S.; Golçalves, D.L.; Pacheco, C.B.; Almeida, W.B.; Thomé, I.P.; Pich, C.T.; Paula, M.M.S.; Silva, L.; Angioletto, E.; Fiori, M.A. Bactericidal polyurethane foam mattresses: Microbiological characterization and effectiveness. *Mater. Sci. Eng. C* **2010**, *30*, 705–708. [CrossRef]
12. Moustafa, H.; Youssef, A.M.; Darwish, N.A.; Abou-Kandil, A.I. Eco-friendly polymer composites for green packaging: Future vision and challenges. *Compos. Part. B-Eng.* **2019**, *172*, 16–25. [CrossRef]
13. Ashjari, H.R.; Dorraji, M.S.S.; Fakhrzadeh, V.; Eslami, H.; Rasoulifard, M.H.; Rastgouy-Houjaghan, M.; Gholizadeh, P.; Kafil, H.S. Starch-based polyurethane/CuO nanocomposite foam: Antibacterial effects for infection control. *Int. J. Biol. Macromol.* **2018**, *111*, 1076–1082. [CrossRef]
14. Demirci, F.; Yildirim, K.; Kocer, H.B. Antimicrobial open-cell polyurethane foams with quaternary ammonium salts. *J. Appl. Polym. Sci.* **2018**, *135*, 16–20. [CrossRef]
15. Sienkiewicz, N. Improvements of polyurethane (PU) foam's antibacterial properties and bio-resistance. In thermal insulation and radiation control technologies for buildings. In *Thermal Insulation and Radiation Control Technologies for Buildings*; Green Energy and Technology; Kośny, J., Yarbrough, D.W., Eds.; Springer: Cham, Switzerland, 2022. [CrossRef]
16. Oh, J.K.; Kohli, N.; Zhang, Y.; min, Y.; Jayaraman, A.; Cisneros-Zevallos, L.; Akbulut, M. Nanoporous aerogel as a bacteria repelling hygienic material for healthcare environment. *Nanotechnology* **2016**, *27*, 85705. [CrossRef] [PubMed]
17. Karthika, M.; Hong, C.; Tianduo, L.; Hongjuan, W.; Sabu, T. Super-hydrophobic graphene oxide-azobenzene hybrids for improved hydrophobicity of polyurethane. *Compos. Part. B-Eng.* **2019**, *173*, 106978. [CrossRef]
18. Ma, W.; Higaki Otsuka, Y.H.; Takahara, A. Perfluoropolyether-infused nano-texture: A versatile approach to omniphobic coatings with low hysteresis and high transparency. *Chem. Commun.* **2013**, *49*, 597–599. [CrossRef]
19. Pendurthi, A.; Movafaghi, S.; Wang, W.; Shadman, S.; Yalin, A.P.; Kota, A.K. Fabrication of nanostructured omniphobic and superomniphobic surfaces with inexpensive CO_2 laser engraver. *ACS Appl. Mater. Interfaces* **2017**, *9*, 25656–25661. [CrossRef]
20. Li, X.; Shan, H.; Cao, M.; Li, B. Facile fabrication of omniphobic PVDF composite membrane via a waterborne coating for anti-wetting and anti-fouling membrane distillation. *J. Membr. Sci.* **2019**, *589*, 117262. [CrossRef]
21. Zhang, M.; Yu, J.; Wang, J. Liquid-infused surfaces based on $ZnO/Co_3O_4/SiO_2$ with omniphobicity and durable anti-corrosion properties. *Surf. Coat. Technol.* **2021**, *407*, 126772. [CrossRef]
22. Wu, X.Q.; Wu, X.; Wang, T.Y.; Zhao, L.; Truong, Y.B.; Ng, D.; Zheng, Y.M.; Xie, Z. Omniphobic surface modification of electrospun nanofiber membrane via vapor deposition for enhanced anti-wetting property in membrane distillation. *J. Membr. Sci.* **2020**, *606*, 118075. [CrossRef]
23. Wei, C.; Tang, Y.; Zhang, G.; Zhang, Q.; Zhan, X.; Chen, F. Facile fabrication of highly omniphobic and self-cleaning surfaces based on water mediated fluorinated nanosilica aggregation. *RSC Adv.* **2016**, *6*, 74340–74348. [CrossRef]
24. Stalder, A.F.; Melchior, T.; Müller, M.; Sage, D.; Blu, T.; Unser, M. Low-bond axisymmetric drop shape analysis for surface tension and contact angle measurements of sessile drops. *Colloids Surf. A* **2010**, *364*, 72–81. [CrossRef]
25. Cheng, D.F.; Masheder, B.; Urata, C.; Hozumi, A. Smooth perfluorinated surfaces with different chemical and physical natures: Their unusual dynamic dewetting behavior toward polar and nonpolar liquids. *Langmuir* **2013**, *29*, 11322–11329. [CrossRef]
26. Tada, H.; Nagayama, H. Chemical vapor surface modification of porous glass with fluoroalkyl-functionalized silanes. 2. Resistance to water. *Langmuir* **1995**, *11*, 136–142. [CrossRef]
27. Vinoda, B.; Vinuth, M.; Yadav, D.; Manjanna, J. Photocatalytic degradation of toxic methyl red dye using silica nanoparticles synthesized from rice husk ash. *J. Environ. Anal. Toxicol.* **2015**, *5*, 336. [CrossRef]
28. Wu, G.; Chen, L.; Liu, L. Effects of silanization and silica enrichment of carbon fibers on interfacial properties of methylphenylsilicone resin composites. *Compos. Part A-Appl. Sci. Manuf.* **2017**, *98*, 159–165. [CrossRef]
29. Oh, J.K.; Rapisand, W.; Zhang, M.; Yegin, Y.; min, Y.; Castillo, A.; Cisneros-Zevallos, L.; Akbulut, M. Surface modification of food processing and handling gloves for enhanced food safety and hygiene. *J. Food Eng.* **2016**, *187*, 82–91. [CrossRef]
30. Vasudevan, R.; Kennedy, A.J.; Merritt, M.; Crocker, F.H.; Baney, R.H. Microscale patterned surfaces reduce bacterial fouling-microscopic and theoretical analysis. *Colloids Surf. B* **2014**, *117*, 225–232. [CrossRef]
31. Wenzel, R.N. Surface roughness and contact angle. *J. Phys. Colloids Chem.* **1949**, *53*, 1466–1467. [CrossRef]
32. Oh, J.K.; Lu, X.; min, Y.; Cisneros-Zevallos, L.; Akbulut, M. Bacterially antiadhesive, optically transparent surfaces inspired from rice leaves. *ACS Appl. Mater. Interfaces* **2015**, *7*, 19274–19281. [CrossRef] [PubMed]
33. Gong, L.; Kyriakides, S.; Jang, W.Y. Compressive response of open-cell foams. Part I: Morphology and elastic properties. *Int. J. Solids Struct.* **2005**, *42*, 1355–1379. [CrossRef]

34. Pan, R.; Cai, M.; Liu, W.; Luo, X.; Chen, C.; Zhang, H.; Zhong, M. Extremely high Cassie-Baxter state stability of superhydrophobic surfaces via precisely tunable dual-scale and triple-scale micro–nano structures. *J. Mater. Chem. A* **2019**, *7*, 18050–18062. [CrossRef]
35. Zhu, H.; Guo, Z.; Liu, W. Adhesion behaviors on superhydrophobic surfaces. *Chem. Commun.* **2014**, *50*, 3900–3913. [CrossRef]
36. Liu, Y.; Zhang, Z.; Hu, H.; Hu, H.; Samanta, A.; Wang, Q.; Ding, H. An experimental study to characterize a surface treated with a novel laser surface texturing technique: Water repellency and reduced ice adhesion. *Surf. Coat. Technol.* **2019**, *374*, 634–644. [CrossRef]
37. Oh, J.K.; Yegin, Y.; Yang, F.; Zhang, M.; Li, J.; Huang, S.; Verkhoturov, S.V.; Schweikert, E.A.; Perez-Lewis, K.; Scholar, E.A.; et al. The influence of surface chemistry on the kinetics and thermodynamics of bacterial adhesion. *Sci. Rep.* **2018**, *8*, 17247. [CrossRef]
38. Daffonchio, D.; Thaveesri, J.; Verstraete, W. Contact angle measurement and cell hydrophobicity of granular sludge from upflow anaerobic sludge bed reactors. *Appl. Environ. Microbiol.* **1995**, *61*, 3676–3680. [CrossRef]
39. Sharma, P.K.; Rao, K.H. Analysis of different approaches for evaluation of surface energy of microbial cells by contact angle goniometry. *Adv. Colloid Interface Sci.* **2002**, *98*, 341–463. [CrossRef]
40. Leckband, D.; Israelachvili, J. Intermolecular forces in biology. *Q. Rev. Biophys.* **2001**, *34*, 105–267. [CrossRef]
41. Pan, S.; Kota, A.K.; Mabry, J.M.; Tuteja, A. Superomniphobic surfaces for effective chemical shielding. *J. Am. Chem. Soc.* **2013**, *135*, 578–581. [CrossRef]
42. Tuteja, A.; Choi, W.; Mabry, J.M.; McKinley, G.H.; Cohen, R.E. Robust omniphobic surfaces. *Proc. Natl. Acad. Sci. USA* **2008**, *105*, 18200–18205. [CrossRef] [PubMed]
43. Khalil-Allafi, J.; Amin-Ahmadi, B.; Zare, M. Biocompatibility and corrosion behavior of the shape memory NiTi alloy in the physiological environments simulated with body fluids for medical applications. *Mater. Sci. Eng. C-Mater.* **2010**, *30*, 1112–1117. [CrossRef]
44. Boughton, B.; Hodgson-Jones, I.S.; Mackenna, R.M.B.; Wheatley VRWormall, A. Some observations of the nature, origin and possible function of the squalene and other hydrocarbons of human sebum. *J. Investig. Dermatol.* **1955**, *24*, 179–189. [CrossRef] [PubMed]
45. Mao, J.; Zhu, D.; Hu, Y.; Wen, X.; Han, Z. Influence of alcohols on the optical clearing effect of skin in vitro. *J. Biomed. Opt.* **2008**, *13*, 021104. [CrossRef] [PubMed]
46. Choi, B.; Tsu, L.; Chen, E.; Ishak, T.S.; Iskandar, S.M.; Chess, S.; Nelson, J.S. Case report: Determination of chemical agent optical clearing potential using in vitro human skin. *Lasers Surg. Med.* **2005**, *36*, 72–75. [CrossRef]
47. Lai, J.H.; Liao, E.Y.; Liao, Y.H.; Sun, C.K. Investigating the optical clearing effects of 50% glycerol in ex vivo human skin by harmonic generation microscopy. *Sci. Rep.* **2021**, *11*, 329. [CrossRef]
48. Defloor, T. The effect of position and mattress on interface pressure. *Appl. Nurs. Res.* **2000**, *13*, 2–11. [CrossRef]
49. Lou, C.-W.; Huang, S.-Y.; Huang, C.-H.; Pan, Y.-J.; Yan, R.; Hsieh, C.-T.; Lin, J.-H. Effects of structure design on resilience and acoustic absorption properties of porous flexible-foam based perforated composites. *Fibers Polym.* **2015**, *16*, 2652–2662. [CrossRef]
50. Jean, J.; Michel, T.; Camille, C. Subtraction of the water spectra from the infrared spectrum of saline solutions. *Appl. Spectrosc.* **1998**, *52*, 234–239. [CrossRef]
51. Basiak, E.; Lenart, A.; Debeaufort, F. How glycerol and water contents affect the structural and functional properties of starch-based edible films. *Polymers* **2018**, *10*, 412. [CrossRef]
52. Carriço, C.S.; Fraga, T.; Carvalho, V.E.; Pasa, V.M.D. Polyurethane foams for thermal insulation uses produced from castor oil and crude glycerol biopolyols. *Molecules* **2017**, *22*, 1091. [CrossRef] [PubMed]
53. Zeitler, V.A.; Brown, C.A. The infrared spectra of some Ti-O-Si, Ti-O-Ti and Si-O-Si compounds. *J. Phys. Chem.* **1957**, *61*, 1174–1177. [CrossRef]

Disclaimer/Publisher's Note: The statements, opinions and data contained in all publications are solely those of the individual author(s) and contributor(s) and not of MDPI and/or the editor(s). MDPI and/or the editor(s) disclaim responsibility for any injury to people or property resulting from any ideas, methods, instructions or products referred to in the content.

Article

Novel Engineered Carbon Cloth-Based Self-Cleaning Membrane for High-Efficiency Oil–Water Separation

Nuo Chen [1], Kexin Sun [1], Huicong Liang [1], Bingyan Xu [1], Si Wu [1], Qi Zhang [1], Qiang Han [1], Jinghai Yang [1] and Jihui Lang [1,2,*]

[1] Key Laboratory of Functional Materials Physics and Chemistry of the Ministry of Education, Jilin Normal University, Siping 136000, China
[2] Siping Hongzui University Science Park, Siping 136000, China
* Correspondence: jhlang@jlnu.edu.cn; Tel./Fax: +86-434-3294566

Abstract: A novel engineered carbon cloth (CC)-based self-cleaning membrane containing a Cu:TiO$_2$ and Ag coating has been created via hydrothermal and light deposition methods. The engineered membrane with chrysanthemum morphology has superhydrophilic and underwater superhydrophobic performance. The cooperativity strategy of Cu doping and Ag coating to the TiO$_2$ is found to be critical for engineering the separation efficiency and self-cleaning skill of the CC-based membrane under visible light due to the modulated bandgap structure and surface plasmon resonance. The CC-based membrane has excellent oil–water separation performance when Cu is fixed at 2.5 wt% and the Ag coating reaches a certain amount of 0.003 mol/L AgNO$_3$. The contact angle of underwater oil and the separation efficiency are 156° and 99.76%, respectively. Furthermore, the membrane has such an outstanding self-cleaning ability that the above performance can be nearly completely restored after 30 min of visible light irradiation, and the separation efficiency can still reach 99.65% after 100 cycles. Notably, the membrane with exceptional wear resistance and durability can work in various oil–water mixtures and harsh environments, indicating its potential as a new platform of the industrial-level available membrane in dealing with oily wastewater.

Keywords: CC-based membrane; doping and coating; oil–water separation; self-cleaning ability; oily wastewater treatment

Citation: Chen, N.; Sun, K.; Liang, H.; Xu, B.; Wu, S.; Zhang, Q.; Han, Q.; Yang, J.; Lang, J. Novel Engineered Carbon Cloth-Based Self-Cleaning Membrane for High-Efficiency Oil–Water Separation. *Nanomaterials* 2023, 13, 624. https://doi.org/10.3390/nano13040624

Academic Editors: Maria Vittoria Diamanti, Massimiliano D'Arienzo, Carlo Antonini, Michele Ferrari and Nidal Hilal

Received: 12 January 2023
Revised: 31 January 2023
Accepted: 31 January 2023
Published: 4 February 2023

Copyright: © 2023 by the authors. Licensee MDPI, Basel, Switzerland. This article is an open access article distributed under the terms and conditions of the Creative Commons Attribution (CC BY) license (https://creativecommons.org/licenses/by/4.0/).

1. Introduction

The oil–water mixture as a pollutant with wide sources causes a lot of economic losses and induces a huge hazard to our living environment [1–6]. Some important procedures including gravity separation, coagulation, air flotation and demulsification have been provided for separating oily wastewater [7,8]. These methods, however, have some drawbacks such as low separation efficiency, secondary pollution and high cost [9–11]. Therefore, the practical work for solving the oily wastewater problem is an essential and urgent challenge [12–14]. Superwettability materials, particularly superhydrophilic and underwater superoleophobic materials, have become a popular topic in recent years [15]. This kind of material with low surface energy usually has high separation efficiency. However, due to oil adhesion from constant use, they are readily polluted and obstructed by oil droplets [16–21]. Hence, the investigation of materials with excellent self-cleaning ability has been undertaken, and some materials including CC, metal mesh, filter membrane, etc., have been proposed [22]. Because of its higher flexibility, good mechanical properties and long-term viability, the CC membrane is an excellent choice for oil–water separation [23–26]. As an excellent catalyst material, TiO$_2$ has excellent catalytic performance under ultraviolet (UV) light irradiation and can realize self-cleaning pollutants on the membrane surface during the process of oil–water separation [27–29]. However, its application is limited because it only responds to ultraviolet light (5% of sunlight) [30,31]. Therefore, it has great significance to developing its self-cleaning performance with visible light responsiveness. The strategies of doping and coating are both effective techniques for modifying the optical band gap of the materials in the visible region and can even enhance catalytic performance [32–34]. The transition metal of d-states Cu doping can generate intermediate energy states between the host material's valence and conduction bands, preventing the recombination of electron–hole (e^-/h^+) pairs effectively [35,36]. The stable emission in the visible region may be obtained by Cu doping as well as in the near-IR region for various optical materials. When Ag is encapsulated into the TiO_2 system, either Ag clusters or their cations can serve as active sites in catalytic reactions to construct highly efficient catalysts because of its special effect on the separation of photogenerated e^-/h^+ when exposed to visible light [37–40]. Additionally, the deposition of precious metals shows obvious plasma resonance and broadens the absorption range of visible light. Therefore, it is believed that the cooperativity strategy of Cu doping and Ag coating in the CC@TiO_2 system may induce the desired performance in oil–water separation application.

According to the information presented above, a novel engineered CC-based self-cleaning membrane for highly efficient oil–water separation is constructed by assembling Cu:TiO_2 and Ag coating to modify the CC via hydrothermal and light deposition methods. The synergistic design for the engineered membrane can possess an outstanding oil–water separation capability; additionally, it can realize an excellent self-cleaning ability under visible light irradiation. For practical applications in industry, the harsh environment test of the engineered membrane is also analyzed in depth.

2. Experimental

2.1. Materials and Synthesis

The used CC was created by annealing the cotton fiber supplied by Xitaotao Trading Co., Ltd. Tetrabutyl titanate ($C_{16}H_{36}O_4Ti$), silver nitrate ($AgNO_3$), copper nitrate trihydrate ($Cu(NO_3)_2·3H_2O$), Sudan III ($C_{22}H_{16}N_4O$), methylene blue ($C_{16}H_{18}ClN_3S·3H_2$), 1, 2-dichloroethane ($C_2H_4Cl_2$) and n-hexane (C_8H_{14}) were purchased from Sinopharm Group Chemical Co., Ltd. Beijing Chemical Plant supplied the hydrochloric acid (HCl) and toluene ($C_6H_5CH_3$). For the oil–water test, various oils were prepared to form the oil–water mixtures. Aladdin Industrial Corporation provided petroleum ether (boiling point 90~120 °C). Soybean oil was bought at a nearby supermarket. The experiment employed deionized water as the experimental water, and all chemical reagents were of an analytical degree.

The preparation process of the engineered CC-based membranes was shown as follows. First, the cotton fibers were divided into the same size of 3 cm × 3 cm and washed 3~5 times with ethanol and deionized water, and then the cleaned cotton fiber cloth was dried in an oven at 80 °C for several hours. Second, the dried cotton fiber cloth was carbonized in a nitrogen atmosphere at 500 °C for 2 h, heated at a rate of 5 °C/min. The CC was produced after carbonization and rinsed three times with deionized water and 6 wt% hydrochloric acid, respectively. The washed CC was then dried in a drying oven at 80 °C for 2 h. Third, the cleaned CC was soaked in alcohol for 10 min for further use. Fourth, 20 mL of concentrated hydrochloric acid, 20 mL of deionized water, 2.5% $Cu(NO_3)_2·3H_2O$ (Cu:Ti = 2.5 wt%, the ratio that was the best in our previous report) and 2.5 mL of tetrabutyl titanate were mixed, and then a light-yellow mixed solution was formed. Following that, the hydrothermal reaction was carried out and the above solution as well as the prepared CC were removed into a PTFE reactor at 150 °C for 5 h. After cooling the PTFE reactor to an ambient temperature, the CC-based membrane containing Cu:TiO_2 was cleaned and then dried in an oven at 80 °C. Finally, the above membrane was soaked in 30 mL of 0.001 mol/L, 0.002 mol/L, 0.003 mol/L, 0.004 mol/L, 0.005 mol/L and 0.006 mol/L $AgNO_3$ under ultraviolet light. After cleaning and drying, the novel engineered CC-based membranes with different Ag contents were synthesized. For convenience, these membranes were named CC@Cu:TiO_2@Ag_Y (Y = 0.1, 0.2, 0.3, 0.4, 0.5 and 0.6).

2.2. Characterizations

A contact angle instrument was used to measure the contact angle (CA) of the CC@Cu:TiO$_2$@Ag$_Y$ membranes (JC2000CD, AC 220 V and 50 Hz). The membrane structure was determined using an X-ray diffractometer (XRD, Rigaku-D/max-2500, Tokyo, Japan). The morphology, valence and composition state of the membranes were observed by a scanning electron microscope (SEM, JSM-7800F, Osaka, Japan), an energy dispersive spectrometer (EDS, 51-XMX1112, London, UK) and an X-ray photoelectron spectrometer (XPS, VGESCALAB 250×, Mark II, London, UK).

3. Results and Discussion

The XRD patterns of the CC, CC@TiO$_2$, CC@Cu:TiO$_2$ and CC@Cu:TiO$_2$@Ag$_Y$ (Y = 0.1, 0.2, 0.3, 0.4, 0.5 and 0.6) membranes are shown in Figure 1. Compared to the original CC in the inset, some strong diffraction peaks appear in the patterns of Figure 1a–h. For CC@TiO$_2$ in Figure 1a, the peaks are located at 27.76°, 36.43°, 39.26°, 41.53°, 44.26°, 54.64°, 56.87°, 62.12°, 64.43°, 69.23° and 70.13°, which correspond to the diffraction of rutile TiO$_2$ crystal planes (110), (101), (200), (111), (210), (211), (220), (002), (310), (301) and (112) [41–43]. It means that the rutile TiO$_2$ is effectively formed on the CC. Compared to the CC@TiO$_2$, the Cu-doped membranes in Figure 1 lack any diffraction peaks for Cu ions (b-h), indicating that Cu ions may be doped into the TiO$_2$ matrix. Moreover, the position of the main diffraction peaks of the Cu-doped membranes has a slight shift to a smaller angle compared with the CC@TiO$_2$ in Figure 1a, illustrating that the lattice parameters of the Cu-doped membranes are a little larger than those of the undoped ones. By fitting the XRD data with the least squares method, the lattice parameters of the two membranes for CC@TiO$_2$ and CC@Cu:TiO$_2$ are calculated. The calculated results of the former (a = 4.593 nm, c = 2.959 nm) and the latter (a = 4.613 nm, c = 2.963 nm) indicate that Cu ions have been incorporated into the TiO$_2$ lattice and may have replaced the Ti ion sites due to the larger ionic radius of Cu ions (R(Cu$^+$) = 0.096 nm and R(Cu^{2+}) = 0.072 nm) [44–48]. For the CC@Cu:TiO$_2$@Ag$_Y$ membranes shown in Figure 1c–h, the peaks at 32.24° and 46.27° belong to AgCl, and the peak at 63.83° belongs to Ag [49]. In addition, it is found that the intensity of the peak related to the Ag element is enhanced gradually with the increase in Ag content, indicating the successful coating of the Ag element on the surface of the CC@Cu:TiO$_2$ membrane.

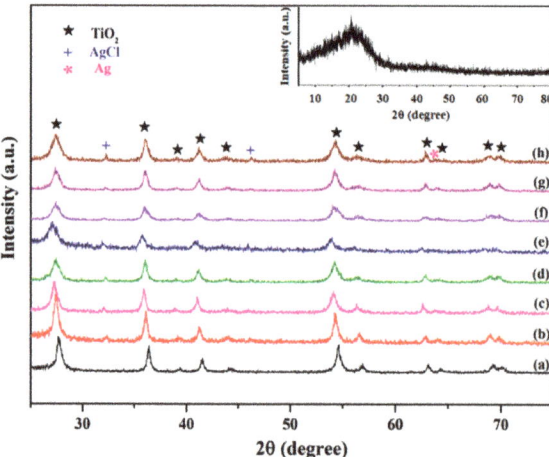

Figure 1. XRD patterns of (a) CC@TiO$_2$, (b) CC@Cu:TiO$_2$ and (c–h) CC@Cu:TiO$_2$@Ag$_Y$ (Y = 0.1, 0.2, 0.3, 0.4, 0.5 and 0.6) membranes, and the inset shows the XRD pattern of CC substrate.

The results of the SEM test are shown in Figure 2 to investigate the morphology of the membranes. In Figure 2a, the original CC is composed of fibers interlaced with each other to form the 3D network structure. When the TiO$_2$ nanomaterials grow on the CC, the original CC surface cannot be seen on the CC@TiO$_2$, indicating that the CC is fully covered by numerous TiO$_2$ nanorods and that these nanorods with diameters of 150~200 nm are arranged closely. Because the original CC fibers are not the planar construction, the nanorods on the CC grow in a flower morphology (the CC@TiO$_2$ membrane in Figure 2b). After Cu doping, it is found that the flower structure is more evident, and a chrysanthemum-like structure on the original basis is formed for the CC@Cu:TiO$_2$ as a result of the Cu-O-Ti in the crystal lattice shown in Figure 2c. The Cu-O-Ti groups hindering the crystal growth are formed during the initial growth process, and then some Ti ion sites might be replaced by the Cu ions to evoke the interfacial energy of the nanorod seeds, which results in the nanorods becoming denser on the flower structure. Therefore, the chrysanthemum-like structure of the membrane is eventually formed with Cu doping in our case [50]. The morphology of the CC@Cu:TiO$_2$@Ag$_Y$ membranes has been slightly modified for additional Ag coating. It is found that some Ag particles with diameters of 200~250 nm are attached to the chrysanthemum surface, and the amount of Ag particles are gradually increased by increasing the Ag content shown in Figure 2d–i.

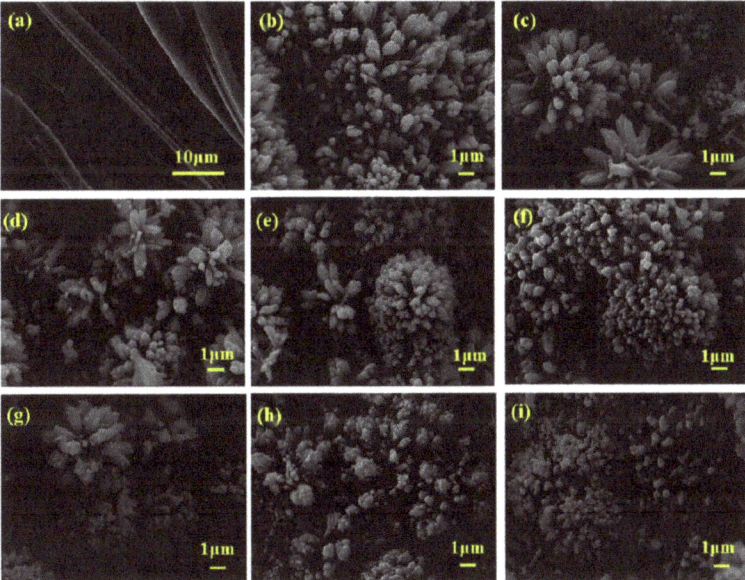

Figure 2. SEM images of (**a**) CC, (**b**) CC@TiO$_2$, (**c**) CC@Cu:TiO$_2$ and (**d–i**) CC@Cu:TiO$_2$@Ag$_Y$ (Y = 0.1, 0.2, 0.3, 0.4, 0.5 and 0.6) membranes.

In Figure 3, the XPS spectrum of the CC@Cu:TiO$_2$@Ag$_{0.3}$ membrane is used to investigate the composition and chemical state of these elements. In the CC@Cu:TiO$_2$@Ag membrane, six elements including Ti, O, C, Cl, Cu and Ag are shown in Figure 3a. The C element comes from the original CC, the Ti and O elements come from TiO$_2$, and Cu(NO$_3$)$_3$H$_2$O and AgNO$_3$ are the sources of the Cu and Ag elements. In Figure 3b of the Ti 2p spectrum, the peak of Ti 2p$_{3/2}$ is located at 458.9 eV and Ti 2p$_{1/2}$ is located at 464.6 eV, which is consistent with the position of Ti^{4+} in the oxide [51]. In Figure 3c of O1s, the peak of O1s can be fitted to O$_{Ti-O}$ at 530.1 eV, O$_{O-H}$ at 532.1 eV and adsorbed O$_2$ at 533.7 eV [52]. The Cu 2p has two peaks at 934.1 eV (Cu 2p$_{3/2}$) and 953.8 eV (Cu 2p$_{1/2}$) in Figure 3d [53], which confirms the existence of Cu^{2+} ions. The peak of Ag 3d in Figure 3e can be fitted to two peaks at 367.8 eV (Ag 3d$_{5/2}$) and 373.9 eV (Ag 3d$_{3/2}$), respectively, and these values indicate that Ag exists in the form of Ag^0 and Ag$^+$ in the CC@Cu:TiO$_2$@Ag$_{0.3}$ membrane [54,55]. Based on the presence of Cl$^-$ in the membrane from the XPS and XRD results, it is assumed that Ag and AgCl are formed from the same Ag source. In our case, the Cl⁻ is thought to be associated with the CC. To demonstrate this, the XPS spectrum of a CC@TiO$_2$ membrane is shown in Figure 3f for comparison. From this spectrum, the four elements of Ti, O, C and Cl appear in the membrane, and it confirms that the Cl⁻ is from the synthesized process of the CC. For researching the distribution of these elements, the EDS mappings of the CC@Cu:TiO$_2$@Ag$_Y$ (Y = 0.1, 0.2, 0.3, 0.4, 0.5 and 0.6) membranes are given in Figure 4. From the mappings, it is found that the elements, especially the Cu and Ag, are all uniformly distributed in these membranes, indicating the uniformity of Cu doping and Ag coating. When these membranes with different Ag contents are compared, it is clear that the Ag content increases as the AgNO$_3$ content increases. It shows that the Ag coating on the membranes gradually increases, which corresponds to the XRD results.

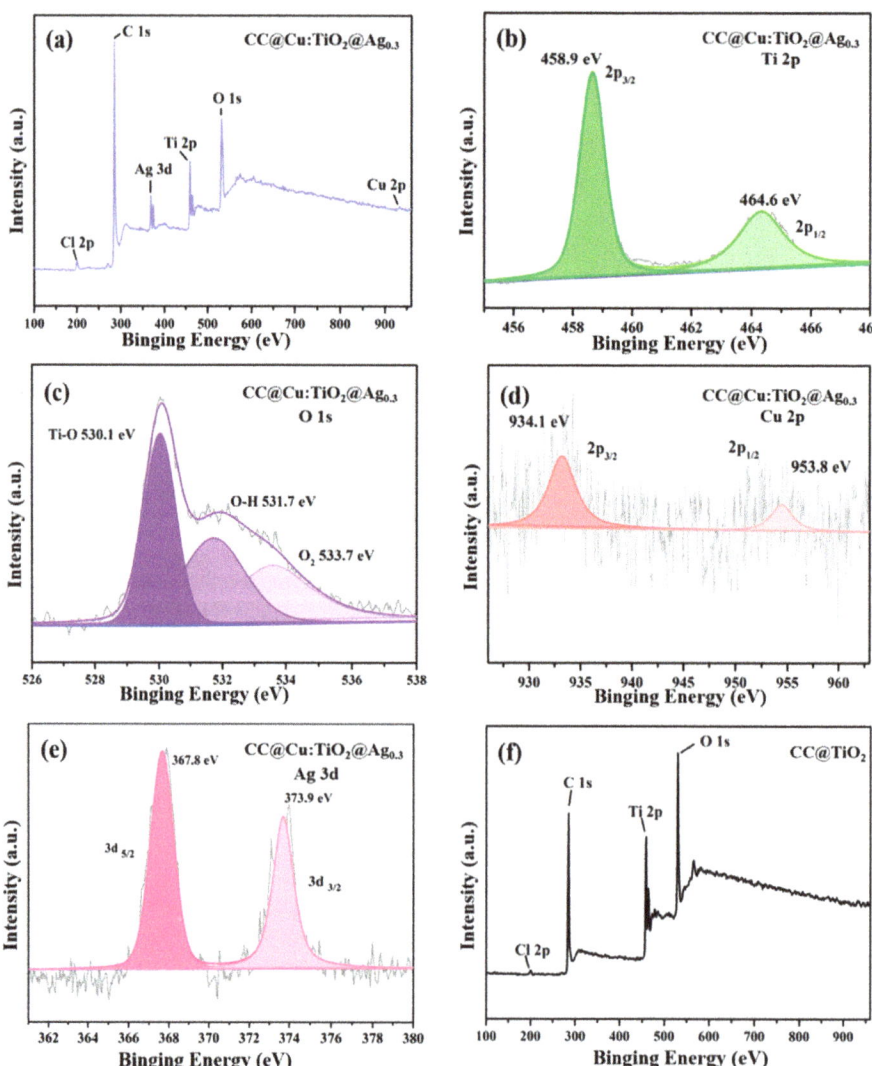

Figure 3. XPS spectra of (**a**–**e**) CC@Cu:TiO$_2$@Ag$_{0.3}$ and (**f**) CC@Cu:TiO$_2$ membranes.

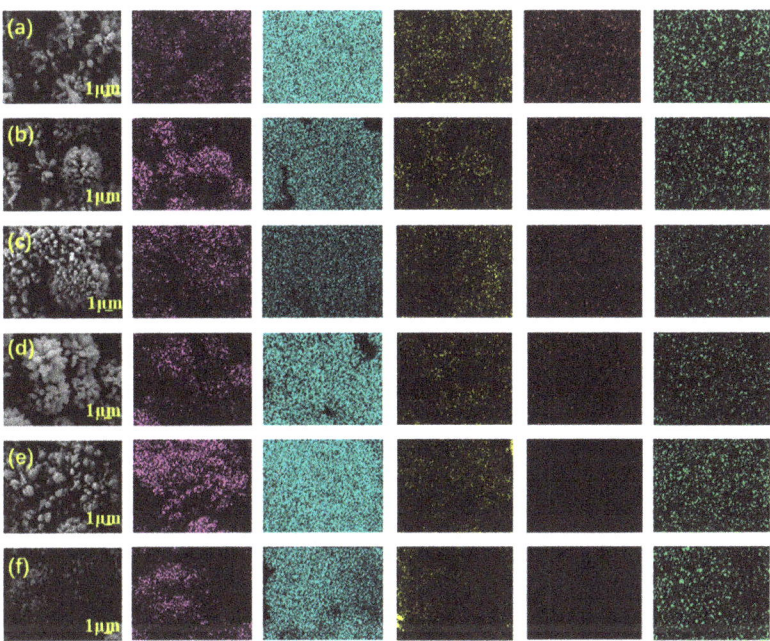

Figure 4. SEM images and the corresponding EDS mappings of (**a**) CC@Cu:TiO$_2$@Ag$_{0.1}$, (**b**) CC@Cu:TiO$_2$@Ag$_{0.2}$, (**c**) CC@Cu:TiO$_2$@Ag$_{0.3}$, (**d**) CC@Cu:TiO$_2$@Ag$_{0.4}$, (**e**) CC@Cu:TiO$_2$@Ag$_{0.5}$, (**f**) CC@Cu:TiO$_2$@Ag$_{0.6}$ membranes.

The underwater oil contact angle determines the membrane's wettability (U-OCA). Compared to the original CC, the wettability of the CC@TiO$_2$, CC@Cu: TiO$_2$ and CC@Cu:TiO$_2$@Ag$_Y$ (Y = 0.1, 0.2, 0.3, 0.4, 0.5 and 0.6) membranes changes significantly. The U-OCA of CC and the CC@TiO$_2$ membranes are 0° and 122.5°, shown in Figure 5, respectively. When doping Cu ions into a TiO$_2$ host, the U-OCA of the CC@Cu:TiO$_2$ increases to 150°. In addition, the U-OCA of the CC@Cu:TiO$_2$@Ag$_Y$ also varies with increasing the Ag content from 0.1 to 0.6, and these values are 150°, 153°, 156°, 151°, 148° and 146°, respectively. Among them, the CC@Cu:TiO$_2$@Ag$_{0.3}$ membrane has the largest U-OCA. Cu doping and Ag coating are thought to take part in promoting the U-OCAs of the membranes. On the one hand, when the membrane is wet, a hydration layer forms on the outside of the membrane due to capillary tension. The water molecules adsorbed on the membrane surface can interact with the membrane's Cu ions to create more hydroxyl radicals. On the other hand, the coating of Ag may mix with the water's electrons to generate OH- and further form the hydroxyl radicals reacting with H$_2$O. These hydroxyl radicals on the CC@Cu:TiO$_2$@Ag$_Y$ membranes will improve membrane hydrophilicity, eventually enhancing membrane oleophobicity. In our case, Cu doping and Ag coating can change the U-OCA. However, the U-OCA of the CC@Cu:TiO$_2$@Ag$_Y$ membranes decreases when the value of Y increases to 0.4 due to the blocked microstructure by superfluous Ag. Figure 6 depicts the hydrophilicity of the CC@Cu:TiO$_2$@Ag$_{0.3}$ membrane in air, the oleophobicity of the membrane in water and its dynamic oleophobic diagram. According to this figure, the water contact angle (WCA) of the CC@Cu:TiO$_2$@Ag$_{0.3}$ membrane in the air is 0°, indicating that water can easily pass through the membrane. That is to say, the membrane has the superhydrophilicity. In addition, an oil droplet can potentially not leave any residue on superoleophobic surfaces even when the oil droplet is deformed and has contacted the membrane surface multiple times, indicating the low oil adhesion of the membranes. The membrane with the lowest oil adhesion is shown by an external force. All of the foregoing investigations show that the constructed CC@Cu:TiO$_2$@Ag$_{0.3}$

membrane has good antifouling performance as well as superhydrophilicity/underwater superoleophobicity, implying that it has a bright future in oil–water separation.

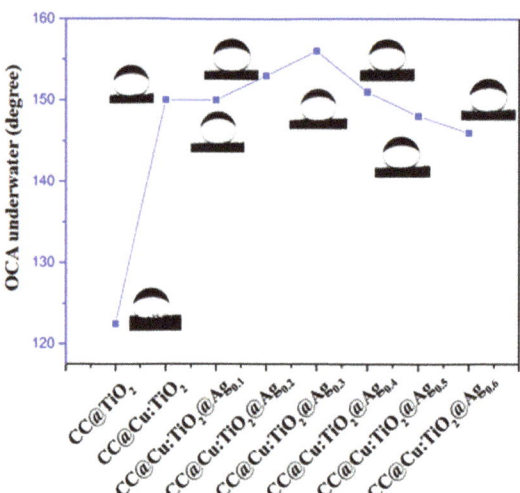

Figure 5. The U-OCA of CC@TiO$_2$, CC@Cu:TiO$_2$ and CC@Cu:TiO$_2$@Ag$_Y$ (Y = 0.1, 0.2, 0.3, 0.4, 0.5 and 0.6) membranes.

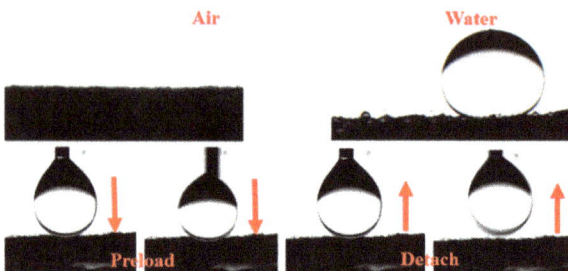

Figure 6. The hydrophilicity of CC@Cu:TiO$_2$@Ag$_{0.3}$ membrane in the air, the oleophobicity of CC@Cu:TiO$_2$@Ag$_{0.3}$ membrane in the water and its dynamic oleophobic diagram.

The oil–water separation test of the membranes is carried out to confirm the initial analysis further and look into the effectiveness of separation and self-cleaning. Figure 7 shows the oil–water separation equipment (a). The tested membranes are clamped onto the two PTFE flange ends with an inner diameter of 18 mm, as illustrated in the oil–water separation device, and the two quartz tubes with a diameter of 15 mm are then placed into the end of the PTFE flange. The produced membranes should be moistened before the separation experiment, and then the mixture of 5 mL of oil and 5 mL of water is poured into the separation device. Briefly, 5 mL of deionized water is colored blue with methylene blue (MB), and 5 mL of oil is colored red with Sudan III for simple observation. When the oil–water mixture is poured into the device, the water barrier generated by the wet membrane prevents the oil from passing through the CC-based membrane. Still, the water can move through the membrane quickly using gravity alone. Figure 7b,c shows that the separation efficiency of the CC@TiO$_2$ membrane without doping and coating is 86.71%, and that of the CC@Cu:TiO$_2$ membrane is boosted to 99.54% due to the larger U-OCA mentioned above, indicating that Cu doping plays a crucial role in oil–water separation efficiency. Moreover, the separation efficiency of the doped and coated membranes of CC@Cu:TiO$_2$@Ag$_Y$ first increases and then decreases as the Ag content increases, and the value reaches 99.76% when the Ag content is fixed at 0.003 mol/L. The water flux of the prepared membranes gradually decreases due to the block gap of the CC when incorporating the Cu and Ag ions. Based on the above analyses, the CC@Cu:TiO$_2$@Ag$_{0.3}$ membrane is thought to have the best oil–water separation. In order to further research the significance of the role of Ag, the long cycle oil–water separation experiment for the CC@Cu:TiO$_2$ and CC@Cu:TiO$_2$@Ag$_{0.3}$ membranes is performed as shown in Figure 8. In this experiment, the membranes are tested for 100 cycles, and every 25 cycles is set as a big cycle. After a big cycle, the membranes are put into 30 mL of deionized water and irradiated with visible light for 15 min and 30 min, respectively. For the first big cycle shown in Figure 8a, the separation efficiency and the water flux for the two membranes both have minor drops from 99.54% to 98.20%, 10524 Lm^{-2}h^{-1} to 10010 Lm^{-2}h^{-1} (CC@Cu:TiO$_2$), 99.76% to 98.77% and 11274 Lm^{-2}h^{-1} to 10422 Lm^{-2}h^{-1} (CC@Cu:TiO$_2$@Ag$_{0.3}$), respectively. The minor drop may be caused by the oil residue attached to the membrane after multiple oil–water separation tests. Before the next big cycle, the membrane is given the self-cleaning treatment at the same irradiation time of 30 min as mentioned above. It is found that the separation efficiency of the CC@Cu:TiO$_2$ and CC@Cu:TiO$_2$@Ag$_{0.3}$ can be restored to 99.32% and 99.65%, and the water flux can be restored to 10530 Lm^{-2}h^{-1} and 11127 Lm^{-2}h^{-1}, respectively. The above values after 100 cycles can even reach 99.12% and 10522 Lm^{-2}h^{-1} (CC@Cu:TiO$_2$), and 99.65% and 11074 Lm^{-2}h^{-1} (CC@Cu:TiO$_2$@Ag$_{0.3}$), demonstrating the superior stability and repeatability of the membranes (Figure 8b). Importantly, the above performance of the CC@Cu:TiO$_2$@Ag$_{0.3}$ membrane after a short irradiation time of 15 min is also much better than that of the CC@Cu:TiO$_2$ membrane under a longer irradiation time of 30 min (Figure 8c). It indicates that the CC@Cu:TiO$_2$@Ag$_{0.3}$ membrane has the best self-cleaning performance that can be restored in a short time due to the Ag coating. Two reasons can account for it, and the self-cleaning mechanism can be seen in Figure 9. For one reason, the membrane's visible light response can be enhanced by incorporating Ag due to the surface plasmon resonance (SPR) effect [56]. Another reason is that the catalytic degradation performance of the membranes can also be improved to some extent due to the adjusted bandgap structure. In the self-cleaning process, the photo-induced photons can be absorbed by the trap sites of Ag, and subsequently, some electrons are produced under the influence of visible light. The electrons from SPR are partially injected into the conduction band of TiO$_2$ because of the higher Fermi energy level of Ag. The growth of the conjugate structure also enables certain electrons in the TiO$_2$ conduction band to transfer quickly onto the CC surface, enhancing electron–hole separation. Moreover, the O$_2$ molecules on the TiO$_2$ surface can be converted into O$_2^-$ by capturing the electrons [57]. Therefore, the degradation ability of pollutants will be effectively enhanced by a certain amount of Ag coating for the CC@Cu:TiO$_2$@Ag$_{0.3}$ membrane. Thus, the CC@Cu:TiO$_2$@Ag$_{0.3}$ membrane has excellent self-cleaning performance, and the Ag coating plays a main role in it.

In order to make the separation environment more suitable for polluted wastewater, the oil–water separation performance of the CC@Cu:TiO$_2$@Ag$_{0.3}$ membrane has been tested by using various oils such as toluene, petroleum ether, soybean oil and 1, 2-dichloroethane. Soybean oil, petroleum ether, toluene and 1, 2-dichloroethane have separation efficiencies and water fluxes of 99.35% and 7020 Lm^{-2}h^{-1}, 99.38% and 7429 Lm^{-2}h^{-1}, 99.51% and 8221 Lm^{-2}h^{-1} and 99.6% and 8502 Lm^{-2}h^{-1}, respectively, as illustrated in Figure 10a. The separation efficiency of various oil–water mixtures with the CC@Cu:TiO$_2$@Ag$_{0.3}$ membrane may vary. The separation efficiencies are all over 99%, indicating that the prepared membrane with the higher oil–water performance perhaps can be used in various oil–water mixture treatments. Figure 10b,c shows the viability tests of the CC@Cu:TiO$_2$@Ag$_{0.3}$ membrane in acidic, alkaline and high-salt environments. The separation efficiency of the CC@Cu:TiO$_2$@Ag$_{0.3}$ membrane in the above environments is above 96% in our case, illustrating that the CC@Cu:TiO$_2$@Ag$_{0.3}$ membrane exhibits the ability to separate an oil–water mixture under harsh acidic, alkaline and high-salt environments of with 2 mol/L.

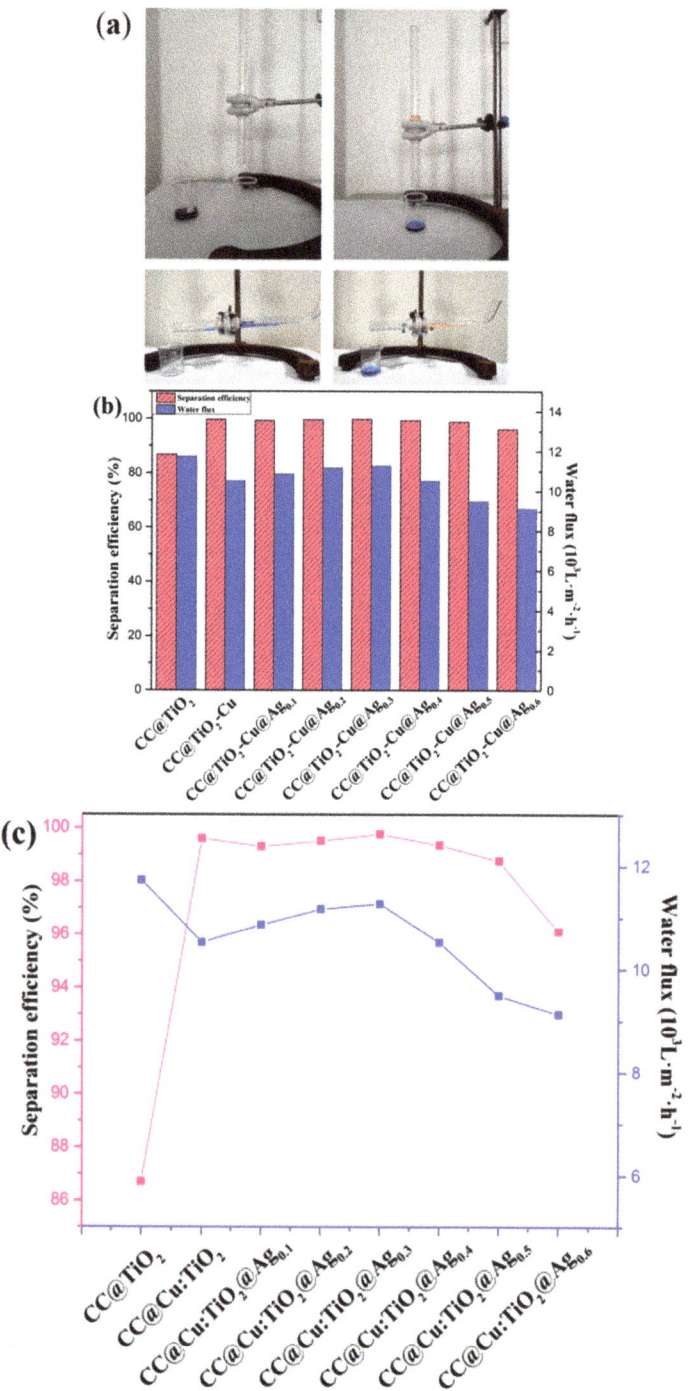

Figure 7. (a) The oil–water separation devices, (b,c) the separation efficiency and the water flux of n-hexane/water mixtures for CC@TiO$_2$, CC@Cu:TiO$_2$ and CC@Cu:TiO$_2$@ Ag$_Y$ (Y = 0.1, 0.2, 0.3, 0.4, 0.5 and 0.6) membranes.

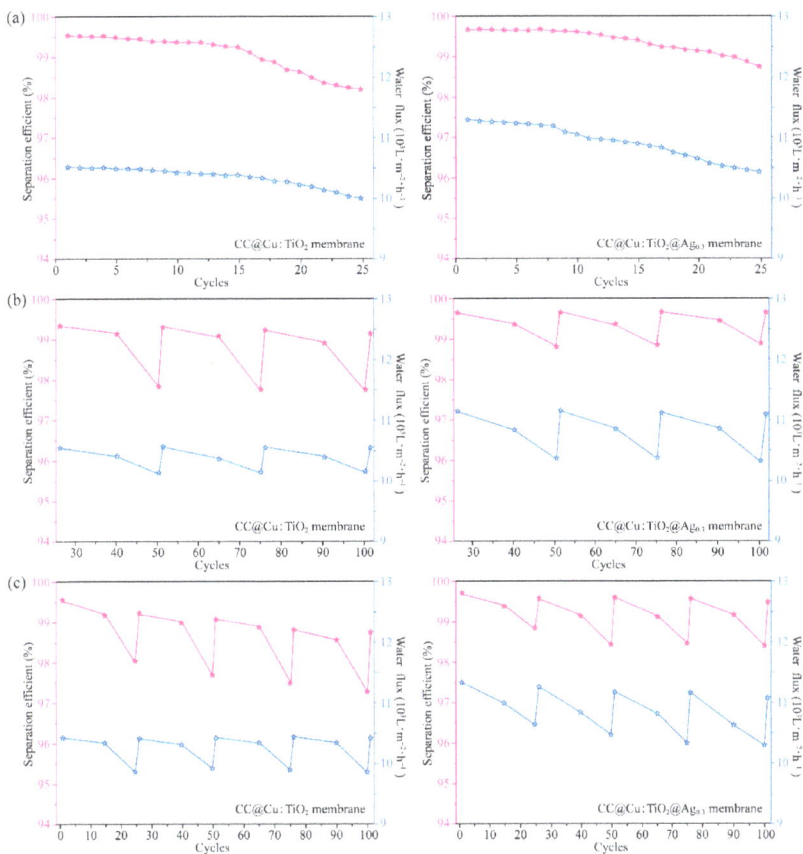

Figure 8. (**a**) The separation efficiency and water flux of the CC@Cu:TiO$_2$ and CC@Cu:TiO$_2$@Ag$_{0.3}$ membranes after 25 cycles, (**b**) the separation efficiency and water flux of CC@Cu:TiO$_2$ and CC@Cu:TiO$_2$@Ag$_{0.3}$ membranes in 75 cycles (irradiation time of 30 min), (**c**) the separation efficiency and water flux of CC@Cu:TiO$_2$ and CC@Cu:TiO$_2$@Ag$_{0.3}$ membranes after 100 cycles (irradiation time of 15 min). The n-hexane/water mixture was used in the experiment.

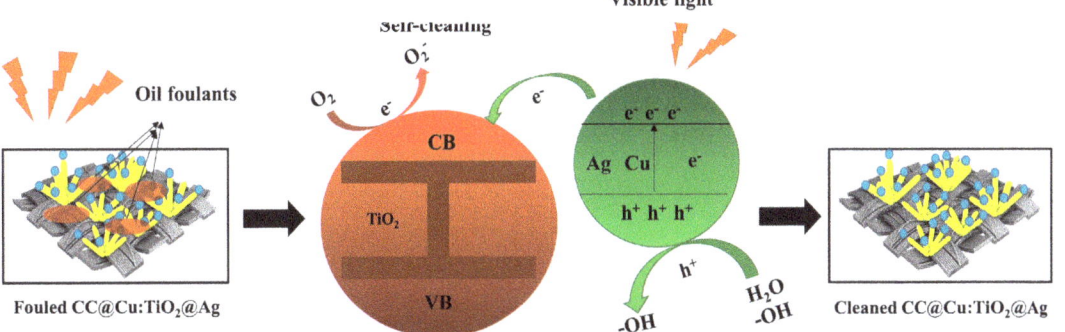

Figure 9. The self-cleaning mechanism of the CC@Cu:TiO$_2$@Ag membrane.

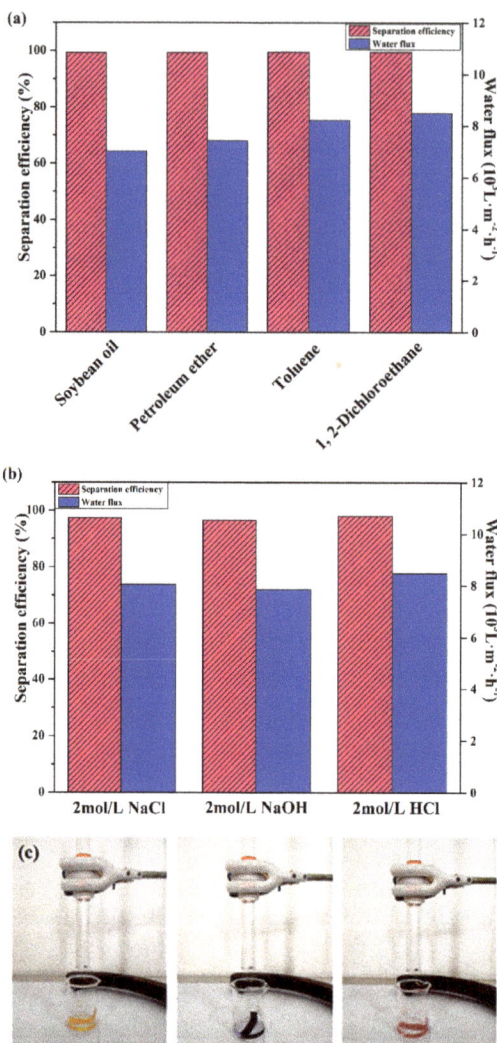

Figure 10. (a) The separation efficiency and the water flux of various oils (soybean oil, petroleum ether, toluene and 1, 2- dichloroethane) for the CC@Cu:TiO$_2$@Ag$_{0.3}$ membrane, (b) The separation efficiency and water flux of acid–base salt for CC@Cu:TiO$_2$@Ag$_{0.3}$ membrane and (c) their separation pictures.

Next, the stability and repeatability tests of the CC@Cu:TiO$_2$@Ag$_{0.3}$ membrane have been conducted. Figure 11 shows the morphology of the CC@Cu:TiO$_2$@Ag$_{0.3}$ membrane before and after all of the oil–water separation experiments mentioned above. Compared with the SEM image before oil–water separation, the overall appearance of the membrane after oil–water separation is unchanged. The original CC is still covered by the Cu:TiO$_2$@Ag structure and a chrysanthemum-like structure can be seen on the original basis. It shows that the prepared membrane not only has an extremely high separation efficiency but also has excellent stability and repeatability. Folding and abrasion tests have also been carried out to prove the mechanical strength of the CC@Cu:TiO$_2$@Ag$_{0.3}$ membrane, shown in Figure 12. It can be observed that the separation efficiency can still reach 99.6% after the membrane is folded 500 times. Notably, when the CC@Cu:TiO$_2$@Ag$_{0.3}$ membrane, attached underneath a weight of 25 g using sandpaper as the base, is pulled for 12 cm and then back to its starting point (every 24 cm named as one cycle), the U-OCA of this membrane is still 153° after abrading for five cycles. It demonstrates that the engineered CC-based membrane has high durability.

Figure 11. (**a**,**b**) SEM images of CC@Cu:TiO$_2$@Ag$_{0.3}$ membrane before oil–water separation at 5000 magnification and 8000 magnification, (**c**,**d**) SEM images of CC@Cu:TiO$_2$@Ag$_{0.3}$ membrane after oil–water separation at 5000 magnification and 8000 magnification.

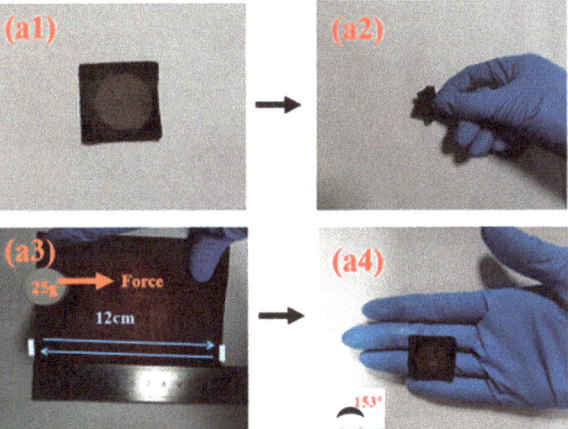

Figure 12. (**a1**,**a2**) The flexibility test and (**a3**,**a4**) durability test of the CC@Cu:TiO$_2$@Ag$_{0.3}$ membrane.

The oil–water separation mechanism is presented to illustrate the separation process. In this experiment, the membrane is first wetted by the water and then 1, 2-dichloroethane (dyed by Sudan III) is poured into the oil–water separation device until the oil droplets penetrate the membrane completely. The intrusion pressure (P_I) is a pressure threshold for the maximum height of an oil column that the membrane can withstand. Through the formula of $P_I = \rho g h_{max}$ (ρ means the density of 1, 2-dichloroethane, and g means the gravitational acceleration) [55], the intrusion pressure is measured on the membrane surface at the maximum allowable liquid height. The maximum liquid column of h_{max} for the membrane is roughly 13.5 cm, as seen in Figure 13; hence, P_I is computed to be 1.663 kPa. When the liquid pressure is less than 1.663 kPa, the oil cannot pass through the membrane.

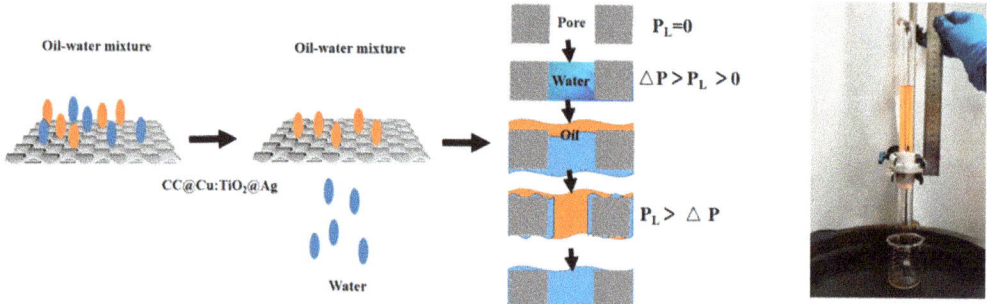

Figure 13. The oil–water separation mechanism of CC@Cu:TiO$_2$@Ag membrane and the intrusion pressure of oil.

4. Conclusions

These hydrothermal and photodeposition methods are used to create a novel engineered CC-based membrane. Cu doping and Ag coating have been found to be beneficial for the excellent oil-separation efficiency and self-cleaning ability of membranes with superhydrophilicity/underwater superoleophobicity due to increased hydroxyl radical generation and enhanced visible light responsiveness. The CC@Cu:TiO$_2$@Ag$_{0.3}$ membrane shows the best performance when the underwater oil contact angle is 156° and the separation efficiency is up to 99.76%. Additionally, the membrane's ability to separate oil from water can be almost recovered under visible light driven for 15 min, indicating that the membrane has excellent self-cleaning ability. Importantly, the membrane can be worked with high separation efficiency in various oils and corrosive environments and even has good durability and wear resistance. The advantages of the simple preparation method and the excellent performance mentioned above indicate that the membranes will have a wide range of practical wastewater treatment applications.

Author Contributions: Writing—Original Draft, N.C.; Software, H.L.; Validation, B.X.; Formal analysis, Q.Z.; Investigation, K.S., Q.H. and J.Y.; Date curation, S.W.; Resources, Q.Z.; Writing—Review & Editing, J.L. All authors have read and agreed to the published version of the manuscript.

Funding: This research was funded by [National Natural Science Foundation of China] grant number [22078124, 21776110, 21878119], [Program for the development of Science and Technology of Jilin Province] grant number [20210203014SF], [Program for the creative and innovative talents of Jilin Province] grant number [2020028].

Data Availability Statement: The data presented in this study are available in the article.

Conflicts of Interest: The authors declare no conflict of interest.

References

1. Zhang, G.; Liu, Y.; Chen, C.; Long, L.; He, J.; Tian, D.; Luo, L.; Yang, G.; Zhang, X.; Zhang, Y. MOF-based cotton fabrics with switchable superwettability for oil–water separation. *Chem. Eng. Sci.* **2022**, *256*, 117695. [CrossRef]
2. Zhang, S.; Su, Q.; Yan, J.; Wu, Z.; Tang, L.; Xiao, W.; Wang, L.; Huang, X.; Gao, J. Flexible nanofiber composite membrane with photothermally induced switchable wettability for different oil/water emulsions separation. *Chem. Eng. Sci.* **2022**, *264*, 118175. [CrossRef]
3. Zhang, X.; Wang, J.; Wang, X.; Cai, Z. Facile preparation of hybrid coating-decorated cotton cloth with superoleophobicity in air for efficient light oil/water separation. *Surf. Interfaces* **2022**, *31*, 102033. [CrossRef]
4. Li, C.; Gao, Z.; Qi, X.; Han, X.; Liu, Z. Preparation and research of Mn-TiO$_2$/ Fe membrane with high efficiency light-oil/water emulsion separation. *Surf. Interfaces* **2022**, *31*, 101995. [CrossRef]

5. Kabiri, B.; Norouzbeigi, R.; Velayi, E. Efficient oil/water separation using grass-like nano-cobalt oxide bioinspired dual-structured coated mesh filters. *Surf. Interfaces* **2022**, *30*, 101825. [CrossRef]
6. Zhang, C.; Gao, J.; Hankett, J.; Varanasi, P.; Borst, J.; Shirazi, Y.; Zhao, S.; Chen, Z. Corn Oil-Water Separation: Interactions of Proteins and Surfactants at Corn Oil/Water Interfaces. *Langmuir* **2020**, *36*, 4044–4054. [CrossRef]
7. Chen, J.; Xiao, X.; Xu, Y.; Liu, J.; Lv, X. Fabrication of hydrophilic and underwater superoleophobic SiO_2/silk fibroin coated mesh for oil/water separation. *J. Environ. Chem. Eng.* **2021**, *9*, 105085. [CrossRef]
8. Bauza, M.; Turnes Palomino, G.; Palomino Cabello, C. MIL-100(Fe)-derived carbon sponge as high-performance material for oil/water separation. *Sep. Purif. Technol.* **2021**, *257*, 117951. [CrossRef]
9. Zhuang, J.; Dai, J.; Ghaffar, S.H.; Yu, Y.; Tian, Q.; Fan, M. Development of highly efficient, renewable and durable alginate composite aerogels for oil/water separation. *Surf. Coat. Technol.* **2020**, *388*, 125551. [CrossRef]
10. Zhao, Y.; Wang, C.; Hu, J.; Li, J.; Wang, Y. Photocatalytic performance of TiO_2 nanotube structure based on TiN coating doped with Ag and Cu. *Ceram. Int.* **2021**, *47*, 7233–7240. [CrossRef]
11. Hu, W.; Huang, J.; Zhang, X.; Zhao, S.; Li, Z.; Zhang, C.; Liu, Y.; Wang, Z. A mechanically robust and reversibly wettable benzoxazine/epoxy/mesoporous TiO_2 coating for oil/water separation. *Appl. Surf. Sci.* **2020**, *507*, 145168. [CrossRef]
12. Bao, Z.; Chen, D.; Li, N.; Xu, Q.; Li, H.; He, J.; Lu, J. Superamphiphilic and underwater superoleophobic membrane for oil/water emulsion separation and organic dye degradation. *J. Mater. Sci.* **2020**, *598*, 117804. [CrossRef]
13. Ren, J.; Tao, F.; Liu, L.; Wang, X.; Cui, Y. A novel TiO_2@stearic acid/chitosan coating with reversible wettability for controllable oil/water and emulsions separation. *Carbohydr. Polym.* **2020**, *232*, 115807. [CrossRef]
14. Cai, Y.; Zhao, Q.; Quan, X.; Feng, W.; Wang, Q. Fluorine-free and hydrophobic hexadecyltrimethoxysilane-TiO_2 coated mesh for gravity-driven oil/water separation. *Colloids Surf. Physicochem. Eng. Asp.* **2020**, *586*, 124189. [CrossRef]
15. Miao, X.; Han, L.; Wang, L.; Wang, M.; Sun, X.; Zhu, X.; Ge, B. Preparation of PDVB/TiO_2 composites and the study on the oil-water separation and degradation performances. *Sci. China Technol. Sc.* **2019**, *62*, 1217–1223. [CrossRef]
16. Chen, Y.; Xie, A.; Cui, J.; Lang, J.; Li, C.; Yan, Y.; Dai, J. One-step facile fabrication of visible light driven antifouling carbon cloth fibers membrane for efficient oil-water separation. *Sep. Purif. Technol.* **2019**, *228*, 115769. [CrossRef]
17. Chen, Z.; Segev, M. Highlighting photonics: Looking into the next decade. *eLight* **2021**, *1*, 2. [CrossRef]
18. Zhang, Z.; Gan, Z.; Bao, R.; Ke, K.; Liu, Z.; Yang, M.; Yang, W. Green and robust superhydrophilic electrospun stereocomplex polylactide membranes: Multifunctional oil/water separation and self-cleaning. *J. Membr. Sci* **2020**, *593*, 117420. [CrossRef]
19. Yang, W.; Shen, H.; Min, H.; Ge, J. Enhanced visible light-driven photodegradation of rhodamine B by Ti^{3+} self-doped TiO_2@Ag nanoparticles prepared using Ti vapor annealing. *J. Mater. Sci.* **2020**, *55*, 701–712. [CrossRef]
20. Hariharan, D.; Thangamuniyandi, P.; Christy, A.J.; Vasantharaja, R.; Selvakumar, P.; Sagadevan, S.; Pugazhendhi, A.; Nehru, L.C. Enhanced photocatalysis and anticancer activity of green hydrothermal synthesized Ag@TiO_2 nanoparticles. *J. Photochem. Photobiol. B Biol.* **2020**, *202*, 111636. [CrossRef]
21. Luo, Z.; Lyu, S.; Mo, D. Cauliflower-like Nickel with Polar Ni(OH)$_2$/NiO$_x$F$_y$ Shell To Decorate Copper Meshes for Efficient Oil/Water Separation. *ACS Omega* **2019**, *4*, 20486–20492. [CrossRef]
22. Cao, Y.; Zhang, W.; Li, B.; Wang, P.; Feng, L.; Wei, Y. Mussel-inspired Ag nanoparticles anchored sponge for oil/water separation and contaminants catalytic reduction. *Sep. Purif. Technol.* **2019**, *225*, 18–23. [CrossRef]
23. Nikitas, P.; Sukosin, T.; Nikolay, I.Z.; Abajo, F.J.G.d. The magnetic response of graphene split-ring metamaterials. *Light-Sci. Appl.* **2013**, *2*, e78. [CrossRef]
24. Lee, D.; So, S.; Hu, G.; Kim, M.; Badloe, T.; Cho, H.; Kim, J.; Kim, H.; Qiu, C.; Rho, J. Hyperbolic metamaterials: Fusing artificial structures to natural 2D materials. *eLight* **2022**, *2*, 1. [CrossRef]
25. Deng, W.; Li, C.; Pan, F.; Li, Y. Efficient oil/water separation by a durable underwater superoleophobic mesh membrane with TiO_2 coating via biomineralization. *Sep. Purif. Technol.* **2019**, *222*, 35–44. [CrossRef]
26. Zhang, Z.; Liu, Z.; Sun, J. Facile preparation of superhydrophilic and underwater superoleophobic mesh for oil/water separation in harsh environments. *J. Dispers. Sci. Technol.* **2019**, *40*, 784–793. [CrossRef]
27. Chen, Y.; Xie, A.; Cui, J.; Lang, J.; Yan, Y.; Li, C.; Dai, J. UV-Driven Antifouling Paper Fiber Membranes for Efficient Oil-Water Separation. *Ind. Eng. Chem. Res.* **2019**, *58*, 5186–5194. [CrossRef]
28. Xie, A.; Cui, J.; Chen, Y.; Lang, J.; Li, C.; Yan, Y.; Dai, J. One-step facile fabrication of sustainable cellulose membrane with superhydrophobicity via a sol-gel strategy for efficient oil/water separation. *Surf. Coat. Technol.* **2019**, *361*, 19–26. [CrossRef]
29. Gautam, J.; Yang, J.; Yang, B.L. Transition metal co-doped TiO_2 nanotubes decorated with Pt nanoparticles on optical fibers as an efficient photocatalyst for the decomposition of hazardous gaseous pollutants. *Colloids Surf. Physicochem. Eng. Asp.* **2022**, *643*, 128786. [CrossRef]
30. Lan, K.; Wang, R.; Zhang, W.; Zhao, Z.; Elzatahry, A.; Zhang, X.; Liu, Y.; Al-Dhayan, D.; Xia, Y.; Zhao, D. Mesoporous TiO_2 Microspheres with Precisely Controlled Crystallites and Architectures. *Chem* **2018**, *4*, 2436–2450. [CrossRef]
31. Chen, Y.; Shen, C.; Wang, J.; Xiao, G.; Luo, G. Green Synthesis of Ag-TiO_2 Supported on Porous Glass with Enhanced Photocatalytic Performance for Oxidative Desulfurization and Removal of Dyes under Visible Light. *ACS Sustain. Chem. Eng.* **2018**, *6*, 13276–13286. [CrossRef]
32. Li, J.; Zhang, Z.; Lang, J.; Wang, J.; Zhang, Q.; Wang, J.; Han, Q.; Yang, J. Tuning red emission and photocatalytic properties of highly active ZnO nanosheets by Eu addition. *J. Lumin.* **2018**, *204*, 573–580. [CrossRef]

33. Zhang, Z.; Song, Y.; Wu, S.; Guo, J.; Zhang, Q.; Wang, J.; Yang, J.; Hua, Z.; Lang, J. Tuning the defects and luminescence of ZnO:(Er, Sm) nanoflakes for application in organic wastewater treatment. *J. Mater. Sci. Mater. Electron.* **2019**, *30*, 15869–15879. [CrossRef]
34. Yang, S.; Yao, J.; Quan, Y.; Hu, M.; Su, R.; Gao, M.; Han, D.; Yang, J. Monitoring the charge-transfer process in a Nd-doped semiconductor based on photoluminescence and SERS technology. *Light-Sci. Appl.* **2020**, *9*, 117. [CrossRef]
35. Qi, P.; Luo, Y.; Shi, B.; Li, W.; Liu, D.; Zheng, L.; Liu, Z.; Hou, Y.; Fang, Z. Phonon scattering and exciton localization: Molding exciton flux in two dimensional disorder energy landscape. *eLight* **2021**, *1*, 6. [CrossRef]
36. Li, Y.; Cao, S.; Zhang, A.; Zhang, C.; Qu, T.; Zhao, Y.; Chen, A. Carbon and nitrogen co-doped bowl-like Au/TiO_2 nanostructures with tunable size for enhanced visible-light-driven photocatalysis. *Appl. Surf. Sci.* **2018**, *445*, 350–358. [CrossRef]
37. Zhang, X.; Zhao, Y.; Mu, S.; Jiang, C.; Song, M.; Fang, Q.; Xue, M.; Qiu, S.; Chen, B. UiO-66-Coated Mesh Membrane with Underwater Superoleophobicity for High-Efficiency Oil-Water Separation. *ACS Appl. Mater. Interfaces* **2018**, *10*, 17301–17308. [CrossRef]
38. Grossmann, T.; Wienhold, T.; Bog, U.; Beck, T.; Friedmann, C.; Kalt, H.; Mappes, T. Polymeric photonic molecule super-mode lasers on silicon. *Light-Sci. Appl.* **2013**, *2*, e82. [CrossRef]
39. Xiong, J.; Wu, S. Planar liquid crystal polarization optics for augmented reality and virtual reality: From fundamentals to applications. *eLight* **2021**, *3*, 3. [CrossRef]
40. Jia, A.; Zhang, Y.; Song, T.; Zhang, Z.; Tang, C.; Hu, Y.; Zheng, W.; Luo, M.; Lu, J.; Huang, W. Crystal-plane effects of anatase TiO_2 on the selective hydrogenation of crotonaldehyde over Ir/TiO_2 catalysts. *J. Catal.* **2021**, *395*, 10–22. [CrossRef]
41. Feng, Z.; Lv, X.; Wang, T. TiO_2 porous ceramic/Ag-AgCl composite for enhanced photocatalytic degradation of dyes under visible light irradiation. *J. Porous Mater.* **2018**, *25*, 189–198. [CrossRef]
42. Kalaiarasi, S.; Sivakumar, A.; Dhas, S.A.M.B.; Jose, M. Shock wave induced anatase to rutile TiO_2 phase transition using pressure driven shock tube. *Mater. Lett.* **2018**, *219*, 72–75. [CrossRef]
43. You, H.; Jin, Y.; Chen, J.; Li, C. Direct coating of a DKGM hydrogel on glass fabric for multifunctional oil-water separation in harsh environments. *Chem. Eng. J.* **2018**, *334*, 2273–2282. [CrossRef]
44. Jin, Z.; Janoschka, D.; Deng, J.; Ge, L.; Dreher, P.; Frank, B.; Hu, G.; Ni, J.; Yang, Y.; Li, J.; et al. Phyllotaxis-inspired nanosieves with multiplexed orbital angular momentum. *eLight* **2021**, *1*, 5. [CrossRef]
45. Shu, Y.; Ji, J.; Xu, Y.; Deng, J.; Huang, H.; He, M.; Leung, D.Y.C.; Wu, M.; Liu, S.; Liu, S.; et al. Promotional role of Mn doping on catalytic oxidation of VOCs over mesoporous TiO_2 under vacuum ultraviolet (VUV) irradiation. *Appl. Catal. B* **2018**, *220*, 78–87. [CrossRef]
46. Chang, X.; Bian, L.; Zhang, J. Large-scale phase retrieval. *eLight* **2021**, *1*, 4. [CrossRef]
47. Stoev, I.D.; Seelbinder, B.; Erben, E.; Maghelli, N.; Kreysing, M. Highly sensitive force measurements in an optically generated, harmonic hydrodynamic trap. *eLight* **2021**, *1*, 7. [CrossRef]
48. Corro, G.; Vidal, E.; Cebada, S.; Pal, U.; Banuelos, F.; Vargas, D.; Guilleminot, E. Electronic state of silver in Ag/SiO_2 and Ag/ZnO catalysts and its effect on diesel particulate matter oxidation: An XPS study. *Appl. Catal. B* **2017**, *216*, 1–10. [CrossRef]
49. Bensouici, F.; Bououdina, M.; Dakhel, A.A.; Tala-Ighil, R.; Tounane, M.; Iratni, A.; Souier, T.; Liu, S.; Cai, W. Optical, structural and photocatalysis properties of Cu-doped TiO_2 thin films. *Appl. Surf. Sci.* **2017**, *395*, 110–116. [CrossRef]
50. Khalid, N.R.; Ahmed, E.; Hong, Z.; Ahmad, M.; Zhang, Y.; Khalid, S. Cu-doped TiO_2 nanoparticles/graphene composites for efficient visible-light photocatalysis. *Ceram. Int.* **2013**, *39*, 7107–7113. [CrossRef]
51. Geng, Z.; Yang, X.; Boo, C.; Zhu, S.; Lu, Y.; Fan, W.; Huo, M.; Elimelech, M.; Yang, X. Self-cleaning anti-fouling hybrid ultrafiltration membranes via side chain grafting of poly(aryl ether sulfone) and titanium dioxide. *J. Mater. Sci.* **2017**, *529*, 1–10. [CrossRef]
52. Mendez-Medrano, M.G.; Kowalska, E.; Lehoux, A.; Herissan, A.; Ohtani, B.; Bahena, D.; Briois, V.; Colbeau-Justin, C.; Rodriguez-Lopez, J.L.; Remita, H. Surface Modification of TiO_2 with Ag Nanoparticles and CuO Nanoclusters for Application in Photocatalysis. *J. Phys. Chem. C* **2016**, *120*, 5143–5154. [CrossRef]
53. Huo, H.; Jiang, Y.; Zhao, T.; Wang, Z.; Hu, Y.; Xu, X.; Lin, K. Quantitatively loaded ultra-small Ag nanoparticles on molecularly imprinted mesoporous silica for highly efficient catalytic reduction process. *J. Mater. Sci.* **2020**, *55*, 1475–1488. [CrossRef]
54. Qi, C.; Zongxue, Y.; Fei, L.; Yang, Y.; Yang, P.; Yixin, P.; Xi, Y.; Guangyong, Z. A novel photocatalytic membrane decorated with RGO-Ag-TiO_2 for dye degradation and oil–water emulsion separation. *J. Chem. Technol. Biotechnol.* **2017**, *93*, 761–775. [CrossRef]
55. Devi, L.G.; Reddy, K.M. Photocatalytic performance of silver TiO_2: Role of electronic energy levels. *Appl. Surf. Sci.* **2011**, *257*, 6821–6828. [CrossRef]
56. Chen, Q.; Yu, J.; Pan, Y.; Zeng, G.; Shi, H.; Yang, X.; Li, F.; Yang, S.; He, Y. Enhancing the photocatalytic and antibacterial property of polyvinylidene fluoride membrane by blending Ag–TiO_2 nanocomposites. *J. Mater. Sci. Mater. Electron.* **2016**, *28*, 3865–3874. [CrossRef]
57. Hou, X.; Hu, Y.; Grinthal, A.; Khan, M.; Aizenberg, J. Liquid-based gating mechanism with tunable multiphase selectivity and antifouling behaviour. *Nature* **2015**, *519*, 70–73. [CrossRef]

Disclaimer/Publisher's Note: The statements, opinions and data contained in all publications are solely those of the individual author(s) and contributor(s) and not of MDPI and/or the editor(s). MDPI and/or the editor(s) disclaim responsibility for any injury to people or property resulting from any ideas, methods, instructions or products referred to in the content.

Article

Benchmarking the Photocatalytic Self-Cleaning Activity of Industrial and Experimental Materials with *ISO 27448:2009*

Hannelore Peeters [1,2], Silvia Lenaerts [1,2] and Sammy W. Verbruggen [1,2,*]

[1] Sustainable Energy, Air & Water Technology (DuEL), Department of Bioscience Engineering, University of Antwerp, Groenenborgerlaan 171, 2020 Antwerp, Belgium
[2] NANOlab Center of Excellence, University of Antwerp, Groenenborgerlaan 171, 2020 Antwerp, Belgium
* Correspondence: sammy.verbruggen@uantwerpen.be

Abstract: Various industrial surface materials are tested for their photocatalytic self-cleaning activity by performing the ISO 27448:2009 method. The samples are pre-activated by UV irradiation, fouled with oleic acid and irradiated by UV light. The degradation of oleic acid over time is monitored by taking water contact angle measurements using a contact angle goniometer. The foulant, oleic acid, is an organic acid that makes the surface more hydrophobic. The water contact angle will thus decrease over time as the photocatalytic material degrades the oleic acid. In this study, we argue that the use of this method is strongly limited to specific types of surface materials, i.e., only those that are hydrophilic and smooth in nature. For more hydrophobic materials, the difference in the water contact angles of a clean surface and a fouled surface is not measurable. Therefore, the photocatalytic self-cleaning activity cannot be established experimentally. Another type of material that cannot be tested by this standard are rough surfaces. For rough surfaces, the water contact angle cannot be measured accurately using a contact angle goniometer as prescribed by the standard. Because of these limitations, many potentially interesting industrial substrates cannot be evaluated. Smooth samples that were treated with an in-house developed hydrophilic titania thin film (PCT/EP2018/079983) showed a great photocatalytic self-cleaning performance according to the ISO standard. Apart from discussing the pros and cons of the current ISO standard, we also stress how to carefully interpret the results and suggest alternative testing solutions.

Keywords: photocatalytic self-cleaning coating; titania thin film; ISO 27448:2009; industrial self-cleaning materials; self-cleaning surfaces

Citation: Peeters, H.; Lenaerts, S.; Verbruggen, S.W. Benchmarking the Photocatalytic Self-Cleaning Activity of Industrial and Experimental Materials with *ISO 27448:2009*. *Materials* **2023**, *16*, 1119. https://doi.org/10.3390/ma16031119

Academic Editors: Maria Vittoria Diamanti, Massimiliano D'Arienzo, Carlo Antonini and Michele Ferrari

Received: 14 December 2022
Revised: 17 January 2023
Accepted: 24 January 2023
Published: 28 January 2023

Copyright: © 2023 by the authors. Licensee MDPI, Basel, Switzerland. This article is an open access article distributed under the terms and conditions of the Creative Commons Attribution (CC BY) license (https://creativecommons.org/licenses/by/4.0/).

1. Introduction

Photocatalytic self-cleaning materials no longer only refer to self-cleaning glass [1–3]. In the construction sector, self-cleaning tiles, wallpaper, paint, window blinds, concrete and asphalt are being produced as well [4–6]. Moreover, self-cleaning coatings are quintessential for solar cells and photovoltaic panels to keep their yield as high as possible [2,7]. Additionally, more and more indoor self-cleaning materials such as self-cleaning paints [8] and fabrics [9,10] are being used. Even photocatalytically self-cleaning membranes for water purification have been made to battle fouling in (waste) water treatment and membrane distillation [11,12]. To put this into numbers, a quick screening of the scientific and patent research is performed. The first photocatalytic self-cleaning applications appeared in the 1990s, eight decades after the photocatalytic degradation reaction was first described [13,14]. The number of scientific papers mentioning photocatalytic self-cleaning surfaces has increased steadily ever since, but it is far outnumbered by the scientific papers on other photocatalytic applications, specifically for water and air treatments [15]. Additionally, the number of patents concerning photocatalytic applications has grown steadily since the 1990s, with a steep increase in annual new patents in 1995–2000 and a levelling off after 2000, possibly indicating the maturation of the technology. Here, the number of photocatalytic water and air treatments heavily outweigh the number of patents on photocatalytic self-cleaning surfaces. On a cumulative basis, self-cleaning surfaces account for about 9% of them, water treatment accounts for 38% of them and air treatment accounts for 53% of the patents for these applications from 1990 to 2007 [15]. For the readers interested in a patent overview, specifically of air treatment patents, we recommend the review by Paz [15], while specific companies and commercial applications are summarised by Mills and Lee [1].

When we are looking more in detail at photocatalytic self-cleaning materials ('*photocatalysis*', '*photocatalytic material*' and '*self-cleaning*') using Google Patents search engine, the oldest patents actually do not mention the term '*self-cleaning*' explicitly in the patent text. The first patent mentioning any self-cleaning action comprises an electrochemical cell with Ti_4O_7 as a photoanode material for the dissociation of water, and it claims the self-cleaning of both the anode and cathode made from a bipolar material by reversing the current (US4422917A, 1981 [16]). With the wide variety of results for these search terms on Google Patents (56,122 in October 2022), of which many appeared to not consider self-cleaning materials, another patent database was used. In the public database Espacenet by the European Patent Office (EPO), 1343 results (in October 2022) were found for the search terms "*self-cleaning*" AND "*photocatalytic*". The oldest documents treat photocatalytic-binder compositions for various applications including self-cleaning paints and waxes and the removal of contaminants from a fluid stream such as air or water, earliest priority 1990 and 1993, published in 1997 and 1995 respectively (US5616532A, WO9511751A1 [17,18]). After these descriptions for binders, more and more patents for (hydrophilic) coatings covering a variety of materials have been published since 1995 [19]. Selected examples include a washing tank, window glass, cover glass, glass/ceramic/vitroceramic-based substrates, a cover for a solar battery, a lamp/luminaire and a handrail [20–26]. The mentioned photocatalyst, when it is specified, is in most cases TiO_2, but also, ZnO and SnO have been used. The more recent patents are more advanced, describing personal apparatus for air disinfection by UV with a photocatalytic self-cleaning coating, including an antiviral effect and a carbon negative self-cleaning inorganic coating [27,28]. The general increasing trend in the number of annually filed patents is not declining thus far, supposing that the 2020–2021 dip can be attributed to the global recession due to the COVID-19 pandemic, as shown in Figure 1.

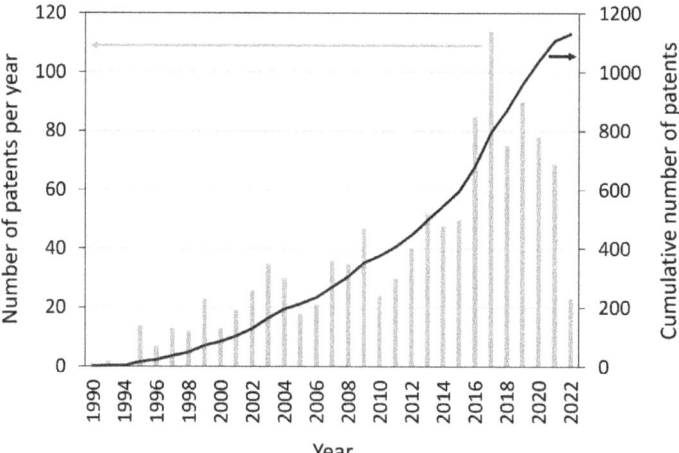

Figure 1. Patents found with Espacenet, EPO's free patent database (worldwise.espace.com), when we were using the search terms "self-cleaning coating" OR "self-cleaning surface") AND "photocatalytic" on 10 October 2022.

The benefits of self-cleaning surfaces speak for themselves. Chemicals/cleaning products, water, energy and labour time are saved. On top of this, the solar energy control can improve, and thus, the energy savings can increase by using self-cleaning surfaces for applications such as photovoltaic panels and smart windows. This is an obvious economic advantage that increases the user's comfort and benefits the environment [2]. The added benefit of photocatalytic self-cleaning materials is that only light is needed to activate the material. After their activation, the materials break down the fouling agents by generating reactive oxygen species (ROS), on the one hand, and make the photocatalytic surface superhydrophilic/superwettable on the other hand. This way both organic foulants (CH-compounds) can be completely degraded to CO_2 and H_2O, while inorganic residues can be easily washed off by the sheeting of water (vide infra) [2]. In order to make the surface photocatalytically active, a photocatalyst is obviously needed. A photocatalyst is typically a semiconductor that is activated by light with sufficient energy to overcome the band gap (E_g). The most commonly used photocatalyst is titanium dioxide (TiO_2), as mentioned earlier, because this semiconductor is affordable, non-toxic, (photo)chemically stable and easily activated. The incident light wavelength needed to activate the semiconductor corresponds to ≤ 388 nm for anatase, with a band gap of about 3.2 eV, implying that (near-)UV light is required for the activation. The large band gap enables TiO_2 to effectively perform many redox reactions, but it is also its largest shortcoming since the solar spectrum at the surface of the Earth only exists for up to 5% of the UV light. Given the importance of an efficient photocatalyst, several strategies to improve the photocatalytic activity of TiO_2 have been proposed. For the interested reader, we gladly refer the reader to specialised review articles on TiO_2 photocatalysis (e.g., one by Verbruggen [29]).

Objectively comparing different photocatalytic self-cleaning materials is not an easy task, since the photocatalytic activity depends on various parameters, such as irradiation intensity, catalyst loading, foulant concentration, temperature and relative humidity, etc. Hence, there is a clear need for a standardised protocol. This protocol should homogenise the testing method by setting fixed values for the parameters that influence the photocatalytic efficiency. The International Standard Organisation (ISO), which was founded in London in 1946, is an independent, non-governmental organisation that publishes such international standards [30]. The first ISO standard for photocatalytic materials was published in 2007: ISO 22197-1:2007—Fine ceramics (advanced ceramics, advanced technical ceramics)—Test method for air-purification performance of semiconducting photocatalytic materials—Part 1: Removal of nitric oxide. Soon, other standards for photocatalytic semiconducting materials were published for the removal of other gases (from ISO 22197-2 to -5), for antibacterial activity (ISO 27447:2009), for self-cleaning activity (ISO 27448:2009), for water purification performance (ISO 10676:2010), for photocatalytic activity in aqueous media (ISO 10678:2010), for the UV light source to be used when testing (ISO 10677:2011) and for antifungal, antiviral and antialgal activity (ISO 13125:2013, 18061-1:2014 and 19635:2016 resp.). In recent years, a true increase in the number of standards has taken place with the publishing of seventeen standards in only four years, and there are two more standards under development. The increase in the number of standards takes the use of different light sources and environments into account (e.g., LED lights and indoor lighting environments), as well as newer, safer (revised standards) and more accurate testing methods for different photocatalytic materials (e.g., in situ FTIR spectra analysis with a recirculating air flow photoreactor for building/construction materials). In this article, we will focus on ISO 27448:2009-Fine ceramics (advanced ceramics, advanced technical ceramics)-Test method for self-cleaning performance of semiconducting photocatalytic materials-Measurement of water contact angle. This standard was first published in 2009, and it was last revised and confirmed in 2020. The photocatalytic self-cleaning activity is measured through the hydrophilicity of the self-cleaning material after being fouled with oleic acid and irradiated with UVA light. More details on the protocol are stated in Section 2. In this study, it is our intention to use the ISO 27448:2009 standard to compare the photocatalytic self-cleaning activity of a variety of industrial self-cleaning surfaces, hygienic materials and materials with an experimental self-cleaning coating. By doing so, we will reveal the benefits and the (unfortunately many) limitations of the standard, highlight the importance of data interpretation, and provide alternative testing procedures.

2. Materials and Methods

All samples were tested according to ISO 27448:2009 Test method for self-cleaning performance of semiconducting photocatalytic materials—Measurement of water contact angle [31]. The method describes the testing of five identical samples. Square-shaped samples of 50 mm by 50 mm were used. After a pre-treatment under UVA light for 24 h at an intensity of $E = 2.0$ mW cm^{-2} to eliminate traces of organic fouling from the surface, the samples were fouled with oleic acid using the dip coating method. The samples were submerged in 0.5 v% oleic acid (Sigma-Aldrich, USA, MI, Saint Louis, 90%) in n-heptane (Chem-Lab NV, 99+ %) and withdrawn at a withdrawal speed of 60 cm min^{-1}. After drying them for 15 min in an oven of 70 °C and cooling them down to room temperature (23 ± 5) °C, the samples were ready for testing. The samples were irradiated using a Philips fluorescence S 25W UV-A lamp, $E = 2$ mW cm^{-2} for wavelengths $\lambda < 400$ nm, ad measured using a calibrated spectroradiometer (Avantes Avaspec-3648-USB2). At certain time intervals (t), five water droplets of 4 µL for each sample were measured using a contact angle goniometer (Ossila) starting at $t = 0$ min, before the irradiation. For each droplet, a video of 10 s at 5 frames per second was captured from the moment right before the droplet touched the sample surface, and it was analysed using the Ossila contact angle software. The equilibrated water contact angle of each droplet was calculated from the last ten frames of each video. The mathematical average of the last three measuring points with a relative error of lower than 10% was defined as the final contact angle (θ_f). A simplified scheme of the experimental set up is given in Figure 2. The samples tested in this study made up a wide variety of materials: kitchen countertops (further denoted as 'Surface materials' A and B), kerbs (samples 'SaniCoat' and 'CleanRock' produced by PolySto), titania-treated carpet tiles and luxury vinyl tiles (LVT) (denoted as 'Carpet tile ref', 'Carpet tile treated', 'LVT ref' and 'LVT treated'), roofing material ('Roofing material'), architectural panels ('Architectural panel' A and B) and a self-cleaning coating developed by our research group applied on glass ('DuEL TiO$_2$') and on architectural panels ('DuEL Architectural panels' A and B). The in-house developed self-cleaning coating was patented (PCT/EP2018/079983) [32], and its specifications have been described in detail elsewhere [3]. The commercial materials listed above were provided by a variety of companies from different industries. Since these materials are either still under development, and/or are patented by the respective companies, we are not allowed to report all of the company names, nor are we allowed to reveal the basic characterization data of the samples, other than the experimental results of the ISO test, due to legal and confidentiality issues. Therefore, most of the materials are labelled with a generic name, as summarised in Table 1.

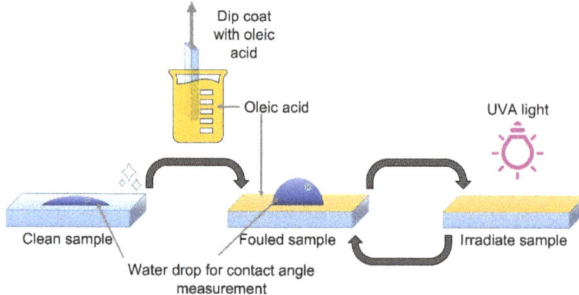

Figure 2. Experimental scheme of ISO 27448:2009 where the water contact angle of a clean sample is measured, the sample is fouled with oleic acid via dip coating, measured again for its water contact angle and irradiated with UV-A light alternatively until the final angle is reached.

Table 1. Overview of the water contact angles of the clean, pre-treated, unfouled surface (θ_{clean}), the oleic-acid-fouled surface before UV irradiation (θ_0), the final contact angle after UV irradiation (θ_f) and irradiation time needed. A dash (-) is used for values that could not be measured.

Sample	θ_{clean} (°)			θ_0 (°)			θ_f (°)			t_f (min)
DuEL TiO$_2$	20	±	7	71	±	20	16	±	4	20
DuEL Architectural panel A	5	±	4	52	±	2	20	±	4	360
DuEL Architectural panel B	3	±	2	52	±	2	23	±	1	3000
SaniCoat	88	±	5	46	±	4	49	±	1	360
CleanRock	71	±	17	64	±	16	50	±	1	420
Architectural panel B	43	±	6	53	±	1	51	±	1	2520
LVT treated	70	±	13	78	±	5	61	±	8	1680
Roofing material	57	±	3	61	±	6	67.7	±	0.5	120
Surface material B	78	±	8	87	±	8	81.0	±	0.4	360
Surface material A	77	±	7	79	±	8	82	±	2	360
LVT ref	93	±	7	87	±	5	87	±	1	1710
Carpet tile treated	-	±	-	-	±	-	-	±	-	-
Carpet tile ref	-	±	-	-	±	-	-	±	-	-

3. Results

As mentioned above, since most of the samples are confidential industrial materials, no characterisation other than the water contact angle measurements were allowed to be performed. The only exception is the in-house developed titania coating (DuEL) on Borofloat® glass, which was fully characterised in our previous study [3]. An example of the evolution of the water contact angle of a typical smooth, hydrophilic, photocatalytic self-cleaning surface is given in Figure 3, with images of a water droplet for the clean, pre-treated, unfouled surface (θ_{clean}), the fouled surface with oleic acid before UV irradiation (θ_0) and the final contact angle after UV irradiation (θ_f).

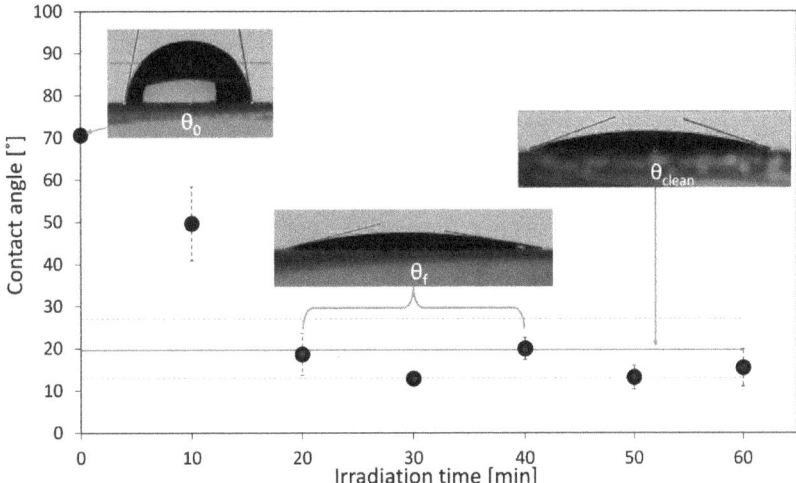

Figure 3. Example of the typical evolution of the water contact angle for a smooth, hydrophilic, photocatalytic self-cleaning surface, in this case, it is DuEL TiO$_2$. The blue line is the clean surface angle with blue dashed lines as error bars. The black dots show the degradation of the deposited oleic acid as the surface regains its hydrophilicity over time.

An overview of the water contact angles of the clean, pre-treated, unfouled surface (θ_{clean}), the fouled surface with oleic acid before UV irradiation (θ_0) and the final contact angle after UV irradiation (θ_f) and irradiation time that needed to be reached θ_f (t_f) are given in Table 1. A dash (-) in the table means that the sample could not be measured according to the guidelines specified by the standard. These cases are discussed in more detail below. Figure 4 shows the evolution from θ_0 to θ_f, and it also summarises t_f for all of the samples.

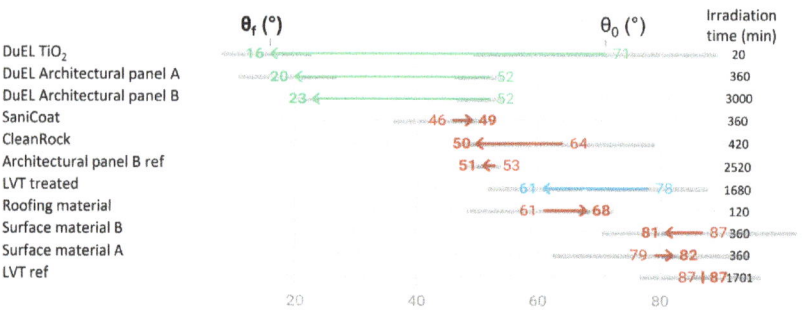

Figure 4. Overview of the evolution the water contact angle from the oleic-acid-fouled surface before UV irradiation (θ_0) to the final contact angle after UV irradiation (θ_f) and the corresponding required irradiation time (t_f) in minutes. The samples can be divided into three groups: the photocatalytically active samples (green arrows), the semi-active samples (blue arrows) and the inactive samples (red arrows). The grey bars represent the experimental error.

It should be clear that the clean contact angle varies greatly depending on the material's surface. The more synthetic materials show rather hydrophobic angles, whilst the samples treated with the DuEL coating (DuEL and DuEL Architectural panels A and B) are characterised by hydrophilic to superhydrophilic angles. The contact angle logically depends on the surface chemistry (e.g., coating) of the investigated material. Yet, there is also a clear noticeable influence of the underlying substrate, as evidenced by the different clean contact angles of DuEL Architectural panel A and B samples and the DuEL TiO$_2$ sample.

In Figure 3, the samples are arranged by their final angles. First, a very important point that needs to be stressed here is that a large final angle does not necessarily imply that the sample is not photocatalytically active, since the water contact angle also strongly depends on the hydrophilicity or -phobicity of the original sample's surface and not only on the remaining amount of oleic acid. Hence, a photocatalytically active surface of a synthetic (hydrophobic) substrate will thus show a larger, hydrophobic, clean and final water contact angle, allowing it to appear that little to no oleic acid has been removed, while in practice, it can simply not be measured. Based on the results of the ISO standard test, the samples can be divided into three groups: the photocatalytically active samples (green), the semi-active samples (blue) and the (seemingly) inactive samples (red). This arbitrary division is based on the difference in the water contact angles before and after irradiation, θ_0 and θ_f, respectively. When a substantial difference is found ($\theta_0/\theta_f > 2$), the sample is categorised as active. When the difference is less significant, but the final angle is still lower than the initial angle and not within the range of error of the measurement ($\theta_0/\theta_f > 1$ and $\theta_0/\theta_f < 2$ with $\theta_f < (\theta_0$—standard deviation [sd])), the sample is categorised as semi-active. When the difference is insignificant ($\theta_0/\theta_f = 1$), the final angle is larger than the initial angle ($\theta_0/\theta_f < 1$) or the final angle falls within the error margin of the initial angle ($\theta_f > (\theta_0$—sd)), the sample is categorised as photocatalytically inactive according to this standard. The three samples categorised as active all have the patented titanium dioxide coating by DuEL. The difference in the photocatalytic activity between these samples might not actually be visible when we are looking at the final angle. Only when we are also comparing the irradiation times needed to reach these final angles, a clearer difference can be seen. The DuEL coating on the Borofloat® glass (DuEL TiO$_2$) required only 20 min to reach the final angle, whereas the DuEL-treated samples, DuEL Architectural panels A and B, required 360 min and 3000 min, respectively, thus the times are one and two orders of magnitude longer. Out of the other samples, only the LVT sample treated with titanium dioxide shows photocatalytic activity, but not to the same extent as the previously mentioned samples. These differences might be due to the influence of the substrate or certain steps in the fabrication process that may have had adverse effects on the photocatalytic properties, e.g., thermal polarisation effects upon heat treatment or the doping of the coating by contaminants from the substrate, leading to excessive charge recombination. Despite the smaller final angles of the SaniCoat and CleanRock samples and the Architectural panel B ref sample compared to that of the treated LVT sample, these samples do not show any photocatalytic activity, as θ_f is within the margin of uncertainty, or it is even larger than θ_0, as is the case for the samples with larger final angles: the Roofing material, the Surface materials A and B and the LVT reference, which can seen on the bottom halves of Table 1 and Figure 4. The samples at the very bottom of the table are the samples that could not be measured, which are represented here with a dash.

Only a few accounts in the literature report on test results obtained with the ISO 27448:2009 method. TiO_2 sol-gel-coated industrial ceramic tiles required 25 or 50 h reach a final angle of below 20° at 2 mW cm^{-2} using the same light intensity as that which was used in this research [33]. Commercial glazed ceramic tiles functionalised with a micrometric TiO_2 layer required 77 h to approach their clean angle of 12.6° at half of the light intensity as that which we used [34]. Transparent TiO_2 and ZnO thin films for polymeric sheets even needed 120 and 180 h of irradiation, respectively, to achieve a final angle of below 5° at 1 mW cm^{-2} [35]. In contrast, the most active DuEL-coated sample in the present study only required 20 min of irradiation to reach the final angle of $(16 \pm 4)°$. Even the example given in the standard protocol itself reports an irradiation time of 70 h before the final angle is reached, clearly confirming the superior performance of our DuEL-coating as a smooth, hydrophilic self-cleaning photocatalytic layer.

4. Discussion

A first and very important observation is that the ISO 27448:2009 method is not applicable to any given sample. Firstly, the method is unsuited to test samples with a very high surface roughness value, as illustrated in Figure 5. Hence, the carpet tiles with a bristle-like surface could simply not be measured using this ISO test, as no water droplets are formed on top of the bristles. Secondly, other samples that are excluded by this standard include water-permeable substrates, highly hydrophobic, powderous or granular materials and visible light-sensitive photocatalysts [31]. For the hydrophobic samples with a contact angle of the clean, pre-treated, unfouled surface (θ_{clean}) that is larger than the initial contact angle (θ_0 after fouling and before UV irradiation), this test will always be inconclusive. In this case, the degradation of oleic acid cannot be measured by the recuperation of hydrophilicity of the material, since the material is not hydrophilic to start with, or at least it is less hydrophilic than the fouled material, which is also visualised in Figure 5. The other disadvantages of this standard are the exclusion of visible light active photocatalysts, and the free choice between the two different methods for applying the oleic acid layer (dip coating versus manual application). Each method comes with a different required irradiance, but without any explication on how or why to choose a given method, which was also pointed out by Mills and Banerjee [36,37]. Another difference between both of the methods of application is that for the dip coating method, there is no intension to quantify the applied amount of oleic acid. The standard does not prescribe measuring a reference sample (e.g., a non-active sample (plain glass) fouled with oleic acid and irradiated in the same way) to account for the possible photodegradation of the oleic acid. Additionally, the volume of the droplet for the water contact angle analysis is not defined, which accounts for that the fact that different articles report different droplet volumes over a range of a few µL. Despite these drawbacks, the standard also has some benefits. Besides the benefits of any standardised protocol to accurately compare the samples under the same conditions, the main benefit of ISO 27448:2009 is its simplicity. The equipment needed for this standard is also not very expensive. An automatic goniometer setup is convenient, and it comes in different price ranges, but nowadays, even smartphone cameras are of high enough quality to take an accurate picture, which can be analysed using freely available software.

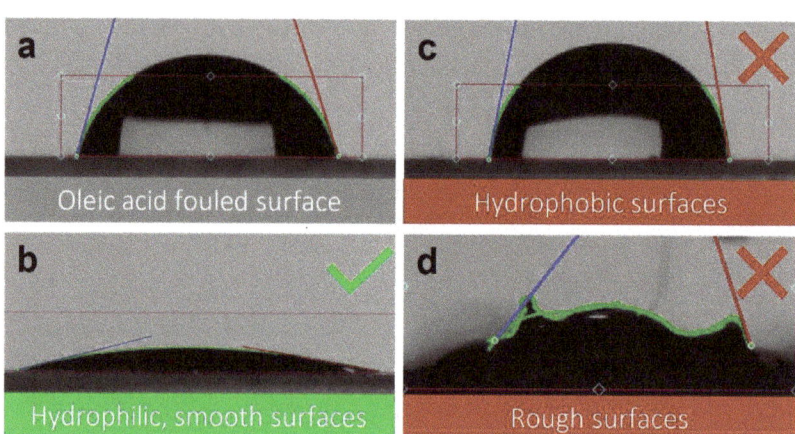

Figure 5. Water droplets on top of different surfaces: a surface fouled with oleic acid (**a**), a clean, smooth and hydrophilic surface (**b**), a clean and hydrophobic surface (**c**) and a clean and rough surface (**d**). For rough surfaces, the water contact angle cannot be measured, and for the hydrophobic surface, the standard cannot be used, which is indicated by the red color and crosses. The green color and checks indicate that hydrophilic, smooth surfaces are suitable for this standard.

We need to stress again here that the fact that because some samples cannot be evaluated using this standard, it does not mean that they are not photocatalytically active. An alternative test for the photocatalytic activity can be performed instead. When we are looking at the available ISO standards, ISO 10678:2010, which tests the photocatalytic activity in an aqueous medium by the degradation of methylene blue, could provide a first alternative. For this test, the initial water contact angle does not matter, and thus, more hydrophobic samples can be measured as well. However, given the high adsorption capacity of the samples with high surface roughness values, this standard does not seem to be the best option for, e.g., the carpet tile samples. The high adsorption capacity can lead to an apparent degradation of methylene blue since this compound is removed from the solution by adsorption onto the surface of the material rather than being degraded photocatalytically. To prevent this, an adsorption period is prescribed by the standard. However, given the extremely high adsorption capacity of the carpet tiles, the adsorption phase would become so long, that the test is no longer practical. Another ISO test that could be performed is ISO 22197:2016 for photocatalytic air purification by the removal of different gaseous compounds. It is important to mention that these standards can be used to indicate the photocatalytic activity of a material, but they are not suited to measure the photocatalytic self-cleaning activity. In the mid-1990s of the previous century, a different method for the evaluation of the photocatalytic self-cleaning activity of glass was proposed by Paz and colleagues [38]. In this method, a thin film of stearic acid, a model compound for organic fouling that is rather similar to oleic acid used in ISO27448, is spin coated onto the self-cleaning samples. Using Fourier Transform Infrared spectroscopy (FTIR), the photoactivity of the samples is calculated by the rate of decrease in the integrated absorbance of C-H stretching vibrations between 2700 and 3000 cm^{-1}. An important condition for this testing method, is that the test samples need to be IR transparent, which unfortunately excludes most of the commercial applications. More recently, another alternative test method has been proposed using smart inks to test the photocatalytic activity of surfaces. In 2009, when ISO 10678:2010, which is used to assess the photocatalytic activity in aqueous media by methylene blue degradation, was under development, Zita, Krýsa and Mills cooperated on the development of an alternative testing method for photocatalytic films. They employed an ink containing the redox dye resazurin as a smart ink that changes color when it is exposed to a photocatalytic reaction. Rather than degrading the dye such as in the case of methylene blue, resazurin changes the color from blue to pink by titania-sensitised reduction. Later on, two more photocatalytic activity indicator inks were introduced: Basic Blue 66 (from blue to colorless) and Acid Violet 7 (from pink to colorless). Given the strong correlation in the photooxidation rates between methylene blue and resazurin oxidation and the high repeatability and reproducibility on top of the faster (less than 10 min), easier and cheaper use of the smart ink test, this (semi-)quantitative method shows great potential. It could boost rapid quality assurance, correct fitting, marketing and in situ assessment to identify 'fake' or photocatalytically inactive products quickly [39]. The method was developed for self-cleaning glass, but also, photocatalytic paints and tiles have been tested [8,40,41]. Since then, several publications have made successful use of this method to measure the photocatalytic activity of self-cleaning materials [39,40,42–45]. In 2018, ISO adapted this method, establishing a new standard ISO 21066:2018 [46].

A second important remark is that a surface that does not show any photocatalytic activity is also not necessarily not self-cleaning. As a matter of fact, there are two main strategies to produce a self-cleaning surface. Either one can make the surface photocatalytically active and hydrophilic, as the materials suitable for the performed ISO standard- or render the surface (super)hydrophobic [2]. The latter method makes use of the water-repelling effect of (super)hydrophobic surfaces, where the formed water beads trap the foulants and carry them off the surface. For this, an inclination angle of at least 10° from the horizontal plane is needed. On top of this, it is harder for the pollutants to stick to the material since the polymers and resins used to make the surface hydrophobic smoothen out the microscopically pitted and pocked surface [2]. A material is categorised as hydrophobic when it has a water contact angle of typically 104° or greater. To be categorised as a hydrophilic material, the water contact angle needs to be below 90°. The different self-cleaning mechanisms are illustrated in Figure 6.

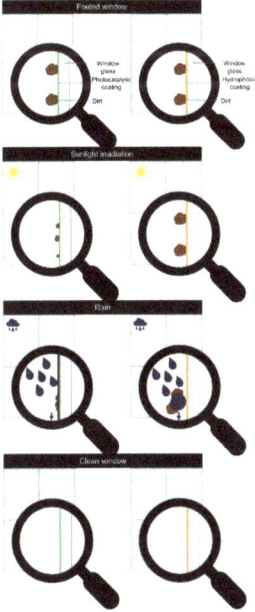

Figure 6. The self-cleaning mechanisms for a photocatalytic, hydrophilic coating on the left and a superhydrophobic coating on the right.

It is equally important to realise that a material that is not self-cleaning is therefore not, by definition, unhygienic. According to a review by Midtdal and Jelle, most of the hydrophobic materials mentioned above actually show a poorer self-cleaning action than the uncoated float glass does. Real-life conditions of outdoor hydrophobic self-cleaning glass hardly ever provide the necessary high impact pressure of water droplets (rain) needed to evacuate the fouling particles. Yet, the hydrophobic coating can reduce and help the cleaning process of the material, since the smoothened surfaces make it harder for foulants to attach to it. These materials can thus be labelled as 'easy-to-clean', rather than 'self-cleaning' [2]. More recently, also, superhydrophobic coatings based on the self-cleaning properties of the Lotus leaf, 'the Lotus Effect', have been developed. To be categorised as superhydrophobic, a surface needs a water contact angle of 150° or more and a roll-off angle of less than 10° [47,48]. Here, rough surface structures and very low surface energies, which are often achieved by silane-based surface modifications, make it hard for the contaminants to adhere to the surface. Instead, the foulants will adhere to water droplets and roll off the surface together. More specifically, the stronger capillary effect compared to the adhesion effect enables contaminant removal. However, water, mostly in the form of rain, is necessary to obtain the self-cleaning activity, as well as the 'easy-to-clean' property. These surfaces can also show anti-icing, anti-corrosion, drag reduction, anti-biofouling, antifogging, self-healing, UV and thermal resistance properties [47,48]. Superhydrophobic self-cleaning surfaces have spread to applications far beyond construction materials, and they are also employed, for example, in display devices [7]. Special attention should be given to the durability of hydrophobic coatings, especially when they are applied by the customer themselves in DIY kits. A lifetime of between 3 and 4 years is expected, whereas a lifetime of 10 years is promised by some producers when the coating is applied by professionals. Photocatalytic self-cleaning coatings are usually applied during the production process, and they are fused with the material, giving them an expected lifetime of up to 30 years for self-cleaning glass, which is the same lifetime as the substrate, e.g., a window itself [2].

5. Conclusions

When one is performing ISO 27448:2009 on a variety of industrial samples and an in-house developed titania coating (DuEL) both on Borofloat® glass and architectural panels, only the DuEL-coated samples show actual photocatalytic self-cleaning activity, according to this standard. The Roofing material treated with titania shows some photocatalytic self-cleaning activity, but the final water contact angle after fouling and irradiation remains large. All of the other tested materials failed to show any activity according to the standard because their surface was either too rough or too hydrophobic. It is our conclusion that ISO 27448:2009 is not necessarily the best test to evaluate the photocatalytic self-cleaning activity of materials. Only a very specific subset of materials is suitable for this test, namely non-powdery, non-granular, hydrophilic materials with surfaces that are smooth enough to support water droplets and the ability for us to perform reliable water contact angle measurements as described by the standard. For substrates that are not hydrophilic, a different test for measuring the photocatalytic activity should be performed. A good candidate test would be the color change measurement using smart inks such as resazurin, Acid Violet 7 or Basic Blue 66. This test can be performed on various substrates and under various conditions, whereas the standard for methylene blue degradation (ISO 10678:2010) evaluates the photocatalytic activity in aqueous media. The proposed method using smart inks is also faster, easier and cheaper for testing the photocatalytic activity than the current ISO standards that are mentioned above are. On the other hand, is it important to mention that a surface that is not photocatalytically active according to the standard is therefore not self-cleaning or unhygienic. While there used to be a lack of proof of the self-cleaning activity of hydrophobic surfaces, recent research into superhydrophobic surfaces might actually achieve a non-photocatalytic self-cleaning function besides self-healing, anti-icing, antibiofouling, anti-corrosion, drag reduction, anti-fogging, UV and thermal and chemical resistance properties, which call the attention of various industries.

The previously mentioned hydrophobic surfaces can be labelled as 'easy-to-clean', and they can reduce the frequency of cleaning procedures. For the complete degradation of organic contaminants, photocatalytic self-cleaning coatings are, for now, still the best option, since these coatings actually mineralise the foulants instead of only rinsing them off upon exposure to water.

Author Contributions: Conceptualization, H.P. and S.W.V.; validation, H.P.; formal analysis, H.P.; investigation, H.P.; resources, H.P. and S.W.V.; data curation, H.P.; writing—original draft preparation, H.P.; writing—review and editing, S.W.V.; visualization, H.P.; supervision, S.L. and S.W.V.; project administration, H.P. and S.W.V.; funding acquisition, H.P., S.L. and S.W.V. All authors have read and agreed to the published version of the manuscript.

Funding: This research was funded by the Research Foundation: Flanders (FWO), through a FWO-SB doctoral fellowship to H.P., grant number: FN 702100002.

Institutional Review Board Statement: Not applicable.

Informed Consent Statement: Not applicable.

Data Availability Statement: The data presented in this study are available on reasonable request from the corresponding author.

Acknowledgments: We kindly acknowledge PolySto and all other companies, who should remain anonymous, for reaching out to us for testing the photocatalytic self-cleaning ability of their materials.

Conflicts of Interest: The funders and the sample providers had no role in the design of the study; in the collection, analyses, or interpretation of data; in the writing of the manuscript; or in the decision to publish the results.

References

1. Mills, A.; Lee, S.-K. A web-based overview of semiconductor photochemistry-based current commercial applications. *J. Photochem. Photobiol. A Chem.* **2002**, *152*, 233–247. [CrossRef]
2. Midtdal, K.; Jelle, B.P. Self-cleaning glazing products: A state-of-the-art review and future research pathways. *Sol. Energy Mater. Sol. Cells* **2013**, *109*, 126–141. [CrossRef]
3. Peeters, H.; Keulemans, M.; Nuyts, G.; Vanmeert, F.; Li, C.; Minjauw, M.; Detavernier, C.; Bals, S.; Lenaerts, S.; Verbruggen, S.W. Plasmonic gold-embedded TiO_2 thin films as photocatalytic self-cleaning coatings. *Appl. Catal. B Environ.* **2020**, *267*, 118654. [CrossRef]
4. Chen, J.; Poon, C.S. Photocatalytic construction and building materials: From fundamentals to applications. *Build. Environ.* **2009**, *44*, 1899–1906. [CrossRef]
5. Smits, M.; Huygh, D.; Craeye, B.; Lenaerts, S. Effect of process parameters on the photocatalytic soot degradation on self-cleaning cementitious materials. *Catal. Today* **2014**, *230*, 250–255. [CrossRef]
6. Mills, A.; Johnston, J.; O'Rourke, C. Photocatalyst Activity Indicator Inks, paiis, for Assessing Self-Cleaning Films. *Acc. Mater. Res.* **2022**, *3*, 67–77. [CrossRef]
7. Adak, D.; Bhattacharyya, R.; Barshilia, H.C. A state-of-the-art review on the multifunctional self-cleaning nanostructured coatings for PV panels, CSP mirrors and related solar devices. *Renew. Sustain. Energy Rev.* **2022**, *159*, 112145. [CrossRef]
8. Baudys, M.; Krýsa, J.; Mills, A. Smart inks as photocatalytic activity indicators of self-cleaning paints. *Catal. Today* **2017**, *280*, 8–13. [CrossRef]
9. Wu, D.; Wang, L.; Song, X.; Tan, Y. Enhancing the visible-light-induced photocatalytic activity of the self-cleaning TiO_2-coated cotton by loading Ag/AgCl nanoparticles. *Thin Solid Films* **2013**, *540*, 36–40. [CrossRef]
10. Moridi Mahdieh, Z.; Shekarriz, S.; Afshar Taromi, F.; Montazer, M. A new method for in situ synthesis of Ag–TiO_2 nanocomposite particles on polyester/cellulose fabric by photoreduction and self-cleaning properties. *Cellulose* **2018**, *25*, 2355–2366. [CrossRef]
11. Zhang, W.; Ding, L.; Luo, J.; Jaffrin, M.Y.; Tang, B. Membrane fouling in photocatalytic membrane reactors (PMRs) for water and wastewater treatment: A critical review. *Chem. Eng. J.* **2016**, *302*, 446–458. [CrossRef]
12. Guo, J.; Yan, Y.S.D.; Lam, F.L.-Y.; Jyoti Deka, B.; Lv, X.; Hau Ng, Y.; Kyoungjin An, A. Self-cleaning BiOBr/Ag photocatalytic membrane for membrane regeneration under visible light in membrane distillation. *Chem. Eng. J.* **2019**, *378*, 122137. [CrossRef]
13. Eibner, A. Über Lichtwirkungen auf Malerfarbstoffe. *Chem. Ztg.* **1911**, *753*, 774–789.
14. Baur, E. Photolysis and electrolysis. *Helv. Chim. Acta* **1918**, *I*, 186. [CrossRef]
15. Paz, Y. Application of TiO_2 photocatalysis for air treatment: Patents' overview. *Appl. Catal. B Environ.* **2010**, *99*, 448–460. [CrossRef]
16. Hayfield, P.C.S. Electrode Material, Electrode and Electrochemical Cell. U.S. Patent 4,422,917A, 18 August 1981.

17. Heller, A.; Pishko, M.; Heller, E. Photocatalyst-Binder Compositions. U.S. Patent 5,616,532A, 21 October 1994.
18. Heller, E.; Heller, A.; Pishko, M. Photocatalyst-Binder Compositions. WO 9511751A1, 21 October 1994.
19. Hayakawa, M.; Kojima, E.; Norimoto, K.; Machida, M.; Kitamura, A.; Watabe, T.; Chikuni, M.; Fujishima, A.; Hashimoto, K. Self-Cleaning Member Having Photocatalytic Hydrophilic Hydrophilic Surface. JP 2002355917A, 10 December 2002.
20. Hasegawa, M.; Uematsu, Y.; Adachi, T.; Harada, W. Washing Tank Having a Self-Cleaning Function. JPH 0938377A, 31 July 1995.
21. Fujishima, A.; Hashimoto, K.; Yada, T.; Miyama, S.; Yoshimoto, T.; Saito, N. Window Glass. JP 3516186B2, 5 April 2004.
22. Fujishima, A.; Hashimoto, K.; Yada, T.; Miyama, S.; Yoshimoto, T.; Saito, N. Cover Glass for Instrument. JP 3224123B2, 29 October 2001.
23. Boire, P.; Talpaert, X. Substrate with a Photocatalytic Coating. U.S. Patent 2002136934A1, 26 August 2002.
24. Hayakawa, M.; Kojima, E.; Watabe, T.; Chikuni, M.; Kitamura, A. Solar Battery with Self-Cleaning Cover. JPH 0983005A, 31 May 1996.
25. Saito, N.; Yotsuyanago, M.; Honda, H.; Iwasaki, B.; Shimizo, K.; Saito, M.; Sakakibara, Y.; Hatakeyama, K.; Taya, A.; Kojima, H.; et al. Photocatalyst, Fluorescent Lamp and Luminaire. JPH 09129012A, 28 May 1995.
26. Hayakawa, M. Self-Cleaning Handrail, and Method of Cleaning Handrail. JP 3774955B2, 17 May 2006.
27. Frorip, A.; Slivinskiy, E.; Taveter, A.; Rennel, A. Personal Apparatus for Air Disinfection by UV. Br. Patent EP 4062948A1, 28 September 2022.
28. Yang, L.; Fang, Y.; Wang, F.; Hu, S.; Liu, Z. Negative Carbon Self-Cleaning Inorganic Coating, Preparation Method Thereof and Obtained Coating. CN 115058132A, 4 June 2022.
29. Verbruggen, S.W. TiO_2 photocatalysis for the degradation of pollutants in gas phase: From morphological design to plasmonic enhancement. *J. Photochem. Photobiol. C Photochem. Rev.* **2015**, *24*, 64–82. [CrossRef]
30. International Organization for Standardization. The ISO Story. Available online: https://www.iso.org/about-us.html (accessed on 29 November 2022).
31. *ISO 27448:2009*; Fine Ceramics (Advanced Ceramics, Advanced Technical Ceramics)-Test Method for Self-Cleaning Performance of Semiconducting Photocatalytic Materials-Measurement of Water Contact Angle. International Standard Organization: Geneva, Switzerland, 2009.
32. Keulemans, M.; Lenaerts, S.; Verbruggen, S.W. Self-Cleaning Coating. International Patent Application No. PCT/EP2018/079983, 9 May 2019.
33. Sciancalepore, C.; Bondioli, F. Durability of SiO_2-TiO_2 photocatalytic coatings on ceramic tiles. *Int. J. Appl. Ceram. Technol.* **2015**, *12*, 679–684. [CrossRef]
34. Tobaldi, D.M.; Graziani, L.; Seabra, M.P.; Hennetier, L.; Ferreira, P.; Quagliarini, E.; Labrincha, J.A. Functionalised exposed building materials: Self-cleaning, photocatalytic and biofouling abilities. *Ceram. Int.* **2017**, *43*, 10316–10325. [CrossRef]
35. Fateh, R.; Dillert, R.; Bahnemann, D. Self-cleaning properties, mechanical stability, and adhesion strength of transparent photocatalytic TiO_2-ZnO coatings on polycarbonate. *ACS Appl. Mater. Interfaces* **2014**, *6*, 2270–2278. [CrossRef]
36. Mills, A.; Hill, C.; Robertson, P. Overview of the current ISO tests for photocatalytic materials. *J. Photochem. Photobiol. A Chem.* **2012**, *237*, 7–23. [CrossRef]
37. Banerjee, S.; Dionysiou, D.D.; Pillai, S.C. Self-cleaning applications of TiO_2 by photo-induced hydrophilicity and photocatalysis. *Appl. Catal. B Environ.* **2015**, *176–177*, 396–428. [CrossRef]
38. Paz, Y.; Luo, Z.; Heller, A. Photooxidative self-cleaning transparent titanium dioxide films on glass. *J. Mater. Res.* **1995**, *10*, 2842–2848. [CrossRef]
39. Zita, J.; Krýsa, J.; Mills, A. Correlation of oxidative and reductive dye bleaching on TiO_2 photocatalyst films. *J. Photochem. Photobiol. A Chem.* **2009**, *203*, 119–124. [CrossRef]
40. Mills, A.; Hepburn, J.; Hazafy, D.; O'Rourke, C.; Krysa, J.; Baudys, M.; Zlamal, M.; Bartkova, H.; Hill, C.E.; Winn, K.R.; et al. A simple, inexpensive method for the rapid testing of the photocatalytic activity of self-cleaning surfaces. *J. Photochem. Photobiol. A Chem.* **2013**, *272*, 18–20. [CrossRef]
41. Mills, A.; Hepburn, J.; Hazafy, D.; O'Rourke, C.; Wells, N.; Krysa, J.; Baudys, M.; Zlamal, M.; Bartkova, H.; Hill, C.E.; et al. Photocatalytic activity indicator inks for probing a wide range of surfaces. *J. Photochem. Photobiol. A Chem.* **2014**, *290*, 63–71. [CrossRef]
42. Mills, A.; McGrady, M. A study of new photocatalyst indicator inks. *J. Photochem. Photobiol. A Chem.* **2008**, *193*, 228–236. [CrossRef]
43. Mills, A.; Hepburn, J.; McFarlane, M. A novel, fast-responding, indicator ink for thin film photocatalytic surfaces. *ACS Appl. Mater. Interfaces* **2009**, *1*, 1163–1165. [CrossRef] [PubMed]
44. Krýsa, J.; Paušová, Š.; Zlámal, M.; Mills, A. Photoactivity assessment of TiO_2 thin films using acid Orange 7 and 4-chlorophenol as model compounds. Part I: Key dependencies. *J. Photochem. Photobiol. A Chem.* **2012**, *25*, 66–71. [CrossRef]
45. Mills, A.; Wells, N.; O'Rourke, C. Correlation between Δabs, ΔrGB (red) and stearic acid destruction rates using commercial self-cleaning glass as the photocatalyst. *Catal. Today* **2014**, *230*, 245–249. [CrossRef]
46. *ISO 21066:2018*; Fine Ceramics (Advanced Ceramics, Advanced Technical Ceramics)—Qualitative and Semiquantitative Assessment of the Photocatalytic Activities of Surfaces by the Reduction of Resazurin in a Deposited Ink Film. International Standard Organization: Geneva, Switzerland, 2018.

47. Zaman Khan, M.; Militky, J.; Petru, M.; Tomková, B.; Ali, A.; Tören, E.; Perveen, S. Recent advances in superhydrophobic surfaces for practical applications: A review. *Eur. Polym. J.* **2022**, *178*, 111481. [CrossRef]
48. Mehmood, U.; Al-Sulaiman, F.A.; Yilbas, B.S.; Salhi, B.; Ahmed, S.H.A.; Hossain, M.K. Superhydrophobic surfaces with antireflection properties for solar applications: A critical review. *Sol. Energy Mater. Sol. Cells* **2016**, *157*, 604–623. [CrossRef]

Disclaimer/Publisher's Note: The statements, opinions and data contained in all publications are solely those of the individual author(s) and contributor(s) and not of MDPI and/or the editor(s). MDPI and/or the editor(s) disclaim responsibility for any injury to people or property resulting from any ideas, methods, instructions or products referred to in the content.

Article

Fabrication of Superhydrophobic Composite Membranes with Honeycomb Porous Structure for Oil/Water Separation

Chunling Zhang [1], Yichen Yang [1], Shuai Luo [1], Chunzu Cheng [2,*], Shuli Wang [2] and Bo Liu [3,*]

[1] School of Materials Science and Engineering, Jilin University, Changchun 130022, China
[2] State Key Laboratory of Bio–Based Fiber Manufacturing Technology, China Textile Academy, Beijing 100025, China
[3] Key Laboratory of Bionic Engineering, Ministry of Education, Jilin University, Changchun 130022, China
* Correspondence: chunzuc@163.com (C.C.); lbb1107@jlu.edu.cn (B.L.)

Abstract: Due to the low separation efficiency and poor separation stability, traditional polymer filtration membranes are prone to be polluted and difficult to reuse in harsh environments. Herein, we reported a nanofibrous membrane with a honeycomb–like pore structure, which was prepared by electrospinning and electrospraying. During the electrospraying process, the addition of poly-dimethylsiloxane and fumed SiO_2 formed pores by electrostatic repulsion between ions, thereby increasing the membrane flux, subsequently reducing the surface energy, and increasing the surface roughness. The results show that when the content of SiO_2 reaches 1.5 wt%, an ultra–high hydrophobic angle (162.1° ± 0.7°) was reached. After 10 cycles of oil–water separation tests of the composite membrane, the oil–water separation flux and separation efficiency was still as high as 5400 L m^{-2} h^{-1} and 99.4%, and the membrane maintained excellent self–cleaning ability.

Keywords: superhydrophobic; oil/water separation; electrospraying; electrospinning

Citation: Zhang, C.; Yang, Y.; Luo, S.; Cheng, C.; Wang, S.; Liu, B. Fabrication of Superhydrophobic Composite Membranes with Honeycomb Porous Structure for Oil/Water Separation. *Coatings* 2022, 12, 1698. https://doi.org/10.3390/coatings12111698

Academic Editor: Maria Vittoria Diamanti

Received: 13 October 2022
Accepted: 3 November 2022
Published: 8 November 2022

Publisher's Note: MDPI stays neutral with regard to jurisdictional claims in published maps and institutional affiliations.

Copyright: © 2022 by the authors. Licensee MDPI, Basel, Switzerland. This article is an open access article distributed under the terms and conditions of the Creative Commons Attribution (CC BY) license (https://creativecommons.org/licenses/by/4.0/).

1. Introduction

A large amount of industrial oily wastewater was discharged everywhere, which has become a serious environmental problem and has endangered our environment [1–3]. Currently, how to treat polluted water efficiently and quickly has become a problem and challenge. Several common treatment methods such as the physical separation method, chemical separation method, biological separation method, and so forth were applied to address this issue [4–6]. However, these methods have complicated operations, make processes cumbersome, show a poor separation performance, and are prone to generating toxic gas and causing secondary pollution [7,8]. Hence, the membrane separation method has become a hot topic in both industry and academia because of its simple operation, low energy consumption, and high separation efficiency [9–12].

Hydrophobic materials are effective in treating oily sewage, therefore the preparation of superhydrophobic surfaces is the most critical step. A superhydrophobic surface is defined as a surface carrying a water contact angle greater than 150° and low contact angle hysteresis [13–15]. The preparation of superhydrophobic surfaces is in two basic principles: (1) Reduce the surface energy by changing the chemical composition of the compound. (2) Reduce the surface energy by changing the surface roughness. Accordingly, how to prepare a hydrophobic surface with low surface energy is the key to solving the problem [16–19].

Recently, PVDF has been widely used in the field of oil–water separation due to its superior mechanical strength and its hydrophobic groups [20–23]. Due to the hydrophobicity of the polyvinylidenefluoride (PVDF) membrane itself, it is easily contaminated when separating oil–in–water emulsions, which reduces the service life of the membrane and increases the cost of oil–water separation [23–25]. Endowing the superhydrophobicity of the film can improve the antifouling performance of the film. In recent years, the blending of inorganic nanoparticles in polymer membranes has attracted attention. It has been demonstrated that the blending of the inorganic filler has led to an increase in membrane permeability and better control of membrane surface properties [26–28]. Nano SiO_2 is a new type of non–toxic, odorless, and pollution–free new ultrafine inorganic material that has a small particle size, high aspect ratio, large specific surface area, and good dispersibility [29]. The abundant hydroxyl groups on the surface enhance the surface effect, have good compatibility with the membrane material and produce a hierarchical structure, which greatly improves the flux and separation efficiency. Yang et al. [24] successfully prepared PVDF rough nanofiber composite membranes by one–step electrospinning technology, with a water contact angle of 135° and an oil–water separation efficiency of 93.9%. Gao et al. [30] used electrospinning and electrostatic spraying to prepare PVDF–SiO_2 nanofiber membranes with surface microsphere structure, the water contact angle reached 152°, and the oil–water separation efficiency was as high as 97%. Although their contact angles are quite high, their flux has a huge drop in 10 oil–water separation cycles, and drops from 7000 to 4000 L m^{-2} h^{-1} after 10 oil–water separation cycle experiments. The reason is that the surface morphology of the composite membrane is composed of electrospinning micropores and microspheres, which are easily polluted and blocked during the oil–water separation process. In order to solve the self–cleaning problem, researchers usually add hydrophobic groups to reduce the surface energy and modify the surface morphology of the membrane [31–34]. Generally, groups containing F and Si elements can reduce the surface energy of the film and increase the hydrophobicity of the film [35–37]. However, to prevent the secondary pollution caused by the F element in the oil–water separation process, a long section of siloxane to the membrane is preferred to achieve the superhydrophobic surface. Polydimethylsiloxane (PDMS) is a fluorine–free polymer. Due to its low surface energy and stable chemical properties, it is suitable for preparing various hydrophobic membranes [38–40]. In this study, we used the hydrosilylation reaction to introduce the rigid group with phenol into the polysiloxane to synthesize a new type of polydimethylsiloxane (DP8) with the long chain segments, and finally prepared a composite film by electrospinning and the electrospray technology with a honeycomb porous structure [41]. In combination with nano SiO_2 and siloxane segments together for modifying the PVDF, superhydrophobicity and self–cleaning were achieved. The oil–water separation performance, stability, and reuse rate of PVDF/DP8/SiO_2 composite membrane for oils of different densities were investigated.

2. Experiments

2.1. Materials

PVDF was purchased from National group chemical reagent Co., Ltd. (Beijing, China). 1,1,3,3–Tetramethyldisiloxane (TMDS) and octamethyl cyclotetrasiloxane (D4) were purchased from Aladdin Technology Co., Ltd (Shanghai, China). The DP8 was synthesized according to our published protocol [41]. Hydrophobic SiO_2 were purchased from Aladdin Reagent Co., Ltd. (Shanghai, China). Methyl blue stain, Sudan III stain, 4,4'-dihydroxydiphenyl, K_2CO_3, bromopropane, methylene chloride, n-hexane, chloroform, carbon tetrachloride, petroleum ether, N,N-dimethylformamide (DMF), tetrahydrofuran (THF), acetone, anhydrous ethanol were obtained from Beijing Chemical Industry (Beijing, China).

2.2. Preparation of Membranes

Synthesis of DP8 is through the silyl–hydrogen reaction of TDMS and D4 to form 2H–PDMS–10, while 4,4'-dihydroxydiphenyl, K_2CO_3 and bromopropene generate intermediate products, then through Claisen rearrangement reaction to generate 3,3'-diallyl-biphenyl-4,4'-diol(DABP), and finally through polymerization reaction to generate long chain polymer, DP8 (Scheme 1) [41]. The PVDF/DP8/SiO_2 membranes were fabricated via simple electrospinning and electrospray technology. Then, 1 g PVDF was dissolved in 10 mL of mixed solvents of DMF and acetone (v/v = 2:3), the blend solutions were subsequently electrospun at a feeding rate of 0.5 mL h^{-1} with 16 kV applied voltage between the working and collecting electrode (receiving distance was set to 12 cm). The synthesized

PVDF nanofiber membrane was placed in a vacuum oven and dried for 24 h, as shown in Scheme 2.

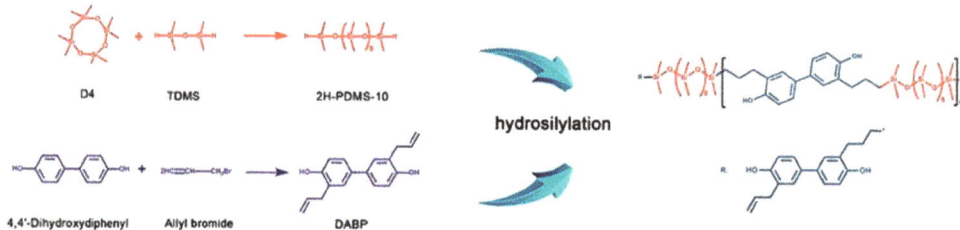

Scheme 1. Synthesis of DP8.

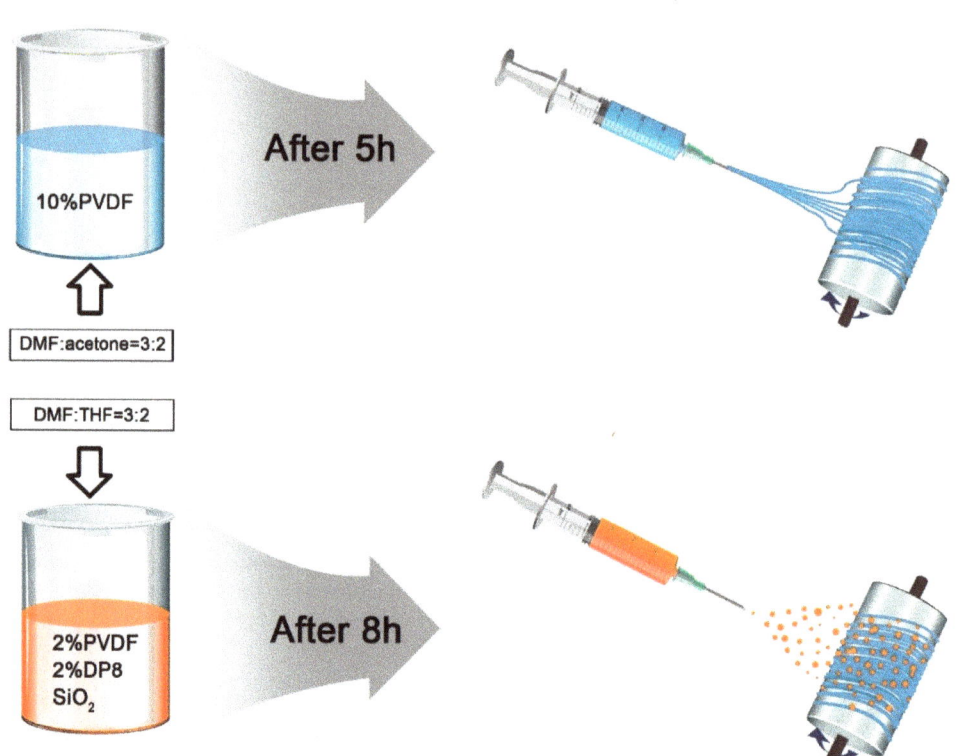

Scheme 2. Electrospinning and electrospraying schematic diagram.

The different weight proportions of SiO_2 (0, 0.5%, 1.0%, 1.5%, 2.5%, 4%), 2 wt% PVDF and 2 wt% DP8, were dissolved in mixed solvents of DMF and THF ($v:v$ = 6:4) and stirred at room temperature until completely dissolved, and then the membrane was fabricated by electrospinning. Here are the parameters of electrospinning: applied voltage of 18 kV, receiving distance of 12 cm, feeding rate of 0.4 mL/h, temperature 20 °C, humidity below 30%. Finally, we took out the double–layer membranes and dried them at 50 °C for 24 h. The obtained membrane was noted as DP8–X, where X presents the concentration of SiO_2 in a mixed solution.

2.3. Characterizations

The surface chemical structure of DP8–X composite membrane was analyzed by Fourier transform infrared (FT–IR) spectra ranging from 4000 to 400 cm^{-1} with a FT–IR system (Nexus 670. Nicolet, WI, USA). Observing the surface morphology of the PVDF/DPn composite membrane with scanning electron microscopy (SEM, FEI XI,30 ESEM FEG). The surface of the sample was sprayed with gold for three minutes, in order to increase the conductivity of the sample. The thermal stability of PVDF/DPn/SiO$_2$ composite membrane analysis was evaluated by Differential Scanning Calorimeter (DSC) (Q20 thermal analyzer, TA, New Castle, PA, USA) and Thermogravimetric measurement (TGA, Perkin–Elmer, Waltham, MA, USA). DSC and TGA experiments were carried out under nitrogen protection, heated with about 3–5 mg DSC sample from 60 to 240 °C at a rate of 10 °C min^{-1}. TGA was heated from 30 to 800 °C (10 °C min^{-1}). Obtained water contact angles were analyzed with a Drop Shape Analyzer (DSA100, Kruss, Hamburg, Germany) by dropping deionized water perpendicularly with a 2 µL syringe onto the PVDF/DP8/SiO$_2$ membrane surface at room temperature. Each membrane was measured five times and the average value recorded.

2.4. Oil–Water Separation

To measure the oil–water separation performance of PVDF/DP8/SiO$_2$ membranes, the PVDF/PD8/SiO$_2$ membrane was placed in the separation device, and the separation device was placed vertically. The effective separation area of the membrane was calculated to be 4.41 cm^2. We poured 60 mL of a mixture of dichloromethane and water ($v:v$ = 1:1) into the glass container above and used gravity as the driving force to separate oil and water. To ensure the separation was complete, the system was maintained for 5–10 min, and two barrels were used to collect oil and water. The equation of separation efficiency was as follows [42,43]:

$$\Phi = V_1/V_2 \tag{1}$$

where Φ is the separation efficiency, and V_1 and V_2 are the volume of oil before and after separation (mL), respectively. The equation of the oil–water separation flux was as follows [44,45]:

$$\text{Flux} = V/A_t \tag{2}$$

where V is the volume of oil phase passing through the membrane (L), A_t is the effective area of separation membrane (m^2), and t is the separation time (h).

3. Results and Discussions

3.1. Preparation of PVDF/DP8/SiO$_2$ Composites

After adding DP8 and SiO$_2$ to the electrospray solution, DP8 and PVDF were randomly distributed in all parts of the PVDF fiber. Figure 1 shows the FTIR spectra of pure PVDF nano-fiber membrane, PVDF/DP8 composite membrane, and the PVDF/DP8/SiO$_2$ nanofiber membranes with different content additives (DP8-X). The PVDF nanofiber membranes show obvious C–H vibration peaks and C–F vibration peaks at 1400 and 1168 cm^{-1}, respectively. When the copolymer DP8 was added, the –CH$_3$ and Si–CH$_3$ bend vibration peaks and Si–O–Si stretching vibration peaks at 2963, 1280, and 1168 cm^{-1} are observed, which indicates that DP8 was successfully added to the PVDF fiber. In the membrane, when SiO$_2$ nanoparticles were added to PVDF/DP8, Si–O stretching vibration peaks became more obvious, and then –CH$_3$, Si–CH$_3$, and C–F bending vibration peaks gradually disappeared because the coated SiO$_2$ nanoparticles on the surface of PVDF/DP8 after the increasing content of SiO$_2$ weakened the intensity of these peaks. Furthermore, this also confirms that SiO$_2$ nanoparticles were successfully doped into PVDF/DP8.

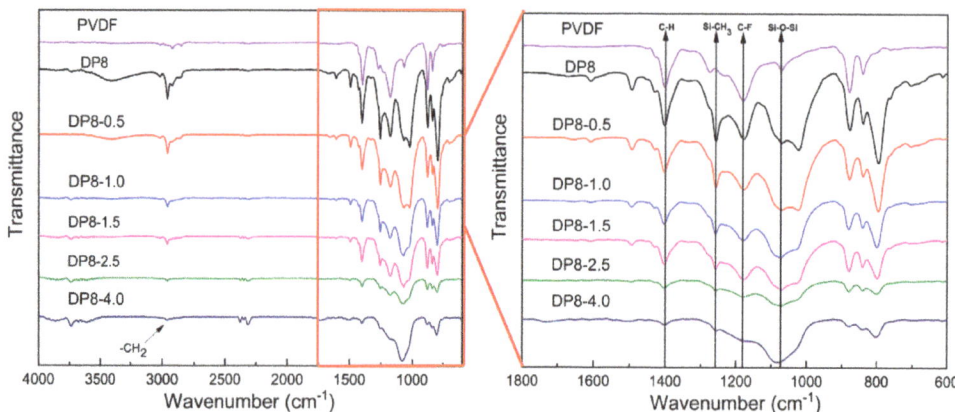

Figure 1. FTIR of pure PVDF nanofiber membrane and DP8–X fiber membrane.

The represented element distribution of PVDF/DP8/SiO$_2$ composite film (DP8–1.5) was analyzed by EDS (Figure 2). The C, O, and F elements were evenly distributed on the surface of the film, and the distribution of Si had a slight reunion. This is because with the increase in the SiO$_2$ content, the phenomenon of reunion between inorganic ions and polymers, and the performance of the membrane are also affected by this phenomenon [30].

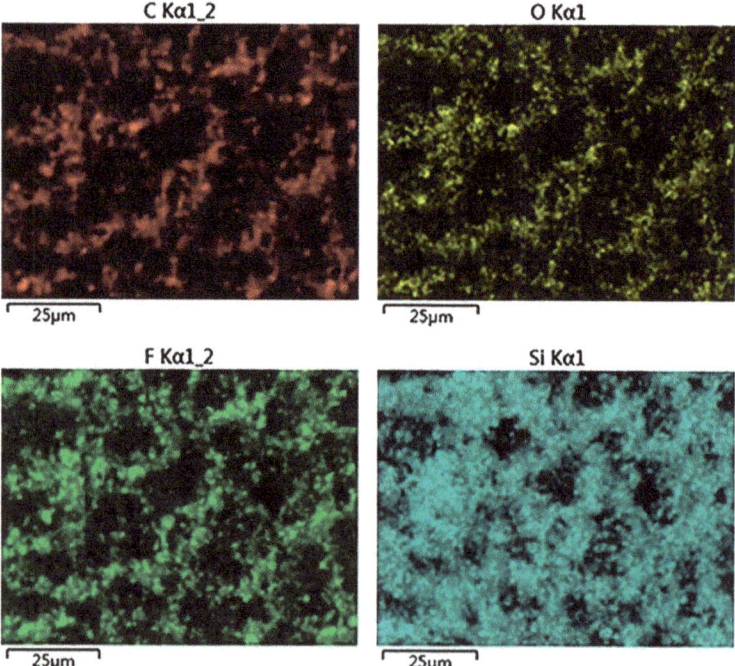

Figure 2. EDS analysis of DP8–1.5 layered membrane: C, O, F, Si element distribution.

3.2. Thermal Performance Analysis of PVDF/DP8/SiO$_2$ Composite Membranes

The thermal stability of the composites was studied by thermogravimetric analysis [46–48]. Thermal stability has always been the key to long–term stable use of oil–water separation membranes [49–51]. Figure 3a shows the thermal weight loss curve of pure PVDF, PVDF/DP8, and DP8–X composite nanofiber membranes. The thermal decomposition process of all membranes was completed in one step, with decomposition beginning at 393 °C and reaching Tmax at 457 °C, and the final Char residues at 800 were 23%. Figure 3b shows the DSC curves of pure PVDF, PVDF/DP8, and DP8–X composite nanofiber membranes. All the films showed a broad endothermic peak at about 160 °C, which is the melting point of the film. After adding SiO_2 nanoparticles, all the endothermic peaks did not move significantly, indicating that the addition of SiO_2 nanoparticles did not reduce the melting point of pure PVDF. Taking the TGA and DSC results together, the addition of SiO_2 nanoparticles and siloxane chain segments did not affect the thermal stability of the membranes.

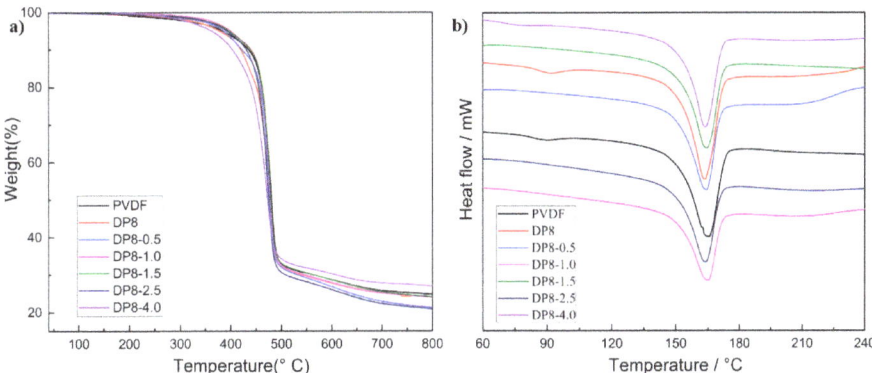

Figure 3. (**a**) TGA and (**b**) DSC curves of pure PVDF nanofiber membrane and DP8–X fiber membranes.

3.3. PVDF/DP8/SiO$_2$ Composite Membrane Water Contact Angle Test and Self–Cleaning Ability Test

Hydrophobicity is an important parameter of oil–water separation performance [52,53]. The water contact angle is tested under dry equilibrium conditions. Figure 4 shows the water contact angles of pure PVDF, PVDF/DP8, and DP8–X composite nanofiber membranes. The water contact angle of the pure PVDF nanofiber membrane was only 121.9° ± 0.7°. However, when the electrospray process fabricated the PVDF/DP8 microspheres on the PVDF nanofiber substrate, the hydrophobic angle of the separation membrane increased from 121.9° ± 0.7° to 145.3° ± 0.8°. This is mainly attributed to the addition of Si-CH$_3$ to reduce its surface energy, and the microsphere structure to increase the surface roughness [36,41], thus improving the hydrophobicity. The hydrophobic angles of DP8–0.5, DP8–1.0, DP8–1.5, and DP8–4.0 were 159.3° ± 2.3°, 159.5° ± 1.7°, 162.1° ± 0.7°, 153.6° ± 2.6°, and 150.4° ± 1.8°, respectively. The hydrophobic angles of the membranes were significantly increased and all the angles reached the values of superhydrophobicity, after incorporating the hydrophobic gas phase SiO$_2$. When the amount of hydrophobic gas phase SiO$_2$ nanoparticles reached up to 4%, the hydrophobic angle was significantly reduced. This is because the interface compatibility between the hydrophobic gas phase SiO$_2$ nanoparticles and PVDF becomes worse, and the agglomeration forms on the surface when the content of SiO$_2$ nanoparticles is too high [43].

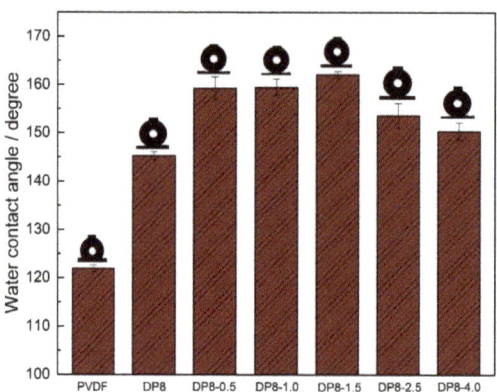

Figure 4. WCAs of pure PVDF nanofiber membrane and DP8–X fiber membranes.

3.4. Oil–Water Separation Test of PVDF/DP8/SiO$_2$ Composite Membrane and Self–Cleaning Ability Test

According to the above analysis, the PVDF/DP8/SiO$_2$ membrane showed excellent hydrophobicity. We used the DP8–1.5 composite membrane for the oil–water separation experiment to further evaluate the oil–water separation performance of the membrane. As shown in Figure 5, 30 mL of water (methyl blue stain) and 30 mL of dichloromethane (Sudan III stain) were configured to form a 60 mL oil–water mixture and placed in a beaker. The DP8–1.5 composite membrane was fixed in the middle of the separation device, and the mixture was slowly poured into the upper funnel. When the oil–water mixture contacted the DP8–1.5 composite membrane, the oil–water mixture was selectively passed through due to the hydrophobic and lipophilic properties of the membrane. Dichloromethane quickly penetrated and passed through the DP8–1.5 superhydrophobic membrane, and finally the oil droplets were collected in the lower beaker. Without any external force, water did not pass through the DP8–1.5 composite membrane, and the separation process was rapid. In order to ensure that all oil droplets could pass through the DP8–1.5 composite membrane, the oil–water separation performance measurement was performed after the entire separation process was maintained for 10 min. Figure 6 shows that the honeycomb porous structure membrane exhibited excellent oil flux and oil–water separation efficiency in the oil–water separation process. The measured oil flux and oil–water separation efficiency could reach up to 5000 L m^{-2} h^{-1} and 99.95%. It is obvious that the flux and separation efficiency was the best when the SiO$_2$ concentration was 1.5%. As shown in Figure 6, the PVDF/DP8/SiO$_2$ composite membrane was used as the separation membrane to carry out 10 cycles of oil–water separation repeatability test. Then the membrane was soaked in absolute ethanol for 5 min and then cleaned and subjected to ultrasonic treatment for 10 min to test the repeatability. Through the above test, it was found that the hydrophobicity of the membrane was not reduced and the superhydrophobic state could be maintained. Within 10 cycles of testing, the oil fluxes of all PVDF/DP8/SiO$_2$ composite membranes remained above 4500 L m^{-2} h^{-1}.

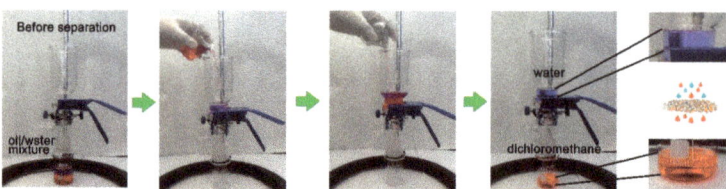

Figure 5. Dichloromethane–water mixture oil–water separation process diagram.

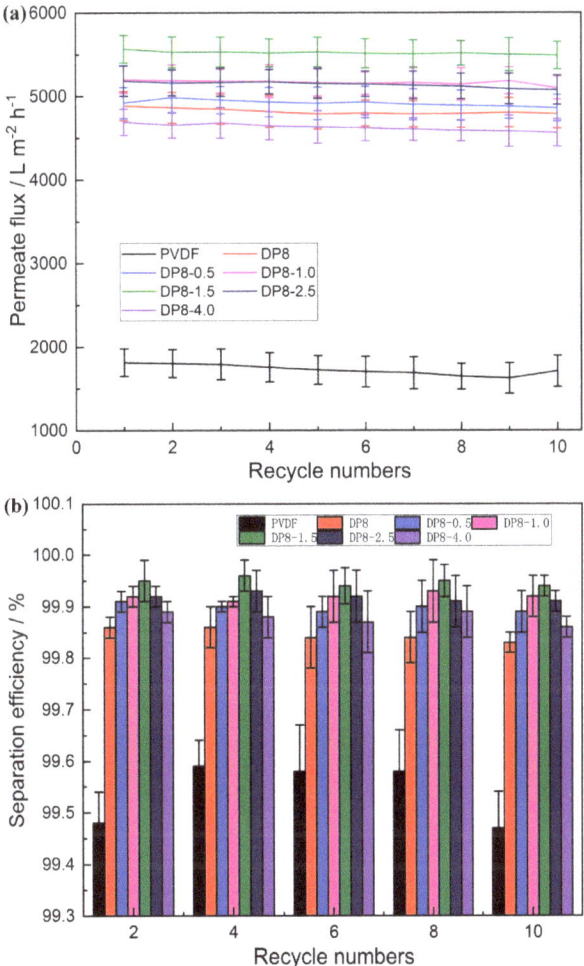

Figure 6. (a) Oil–water separation flux (b) efficiency of PVDF nanofiber membrane and DP–X layered membrane.

In order to study the oil–water separation performance of the DP8–1.5 superhydrophobic composite membrane more comprehensively, five different types of dichloromethane–water, n–hexane–water, chloroform–water, carbon tetrachloride–water and petroleum ether-water were applied for the oil–water mixing system and oil–water separation test. We used the same experimental method for separating the mixture, and the results are shown in Figure 7. When the density of oil was less than the density of water, we tilted the instrument at 45°, similar to n–hexane and petroleum ether. When five different mixtures were separated by the DP8–1.5 honeycomb porous structure membrane, it was found that all the oil fluxes and separation efficiencies were similar. The oil fluxes were 5552, 5498, 5511, 5489, 5545 L m^{-2} h^{-1}, respectively. When the content of SiO$_2$ was 1.5%, it showed a high oil flux to the different oils. The separation efficiency for different oils could be maintained at more than 99.94%. It can be seen that the DP8–1.5 composite membrane had an excellent oil–water separation performance for different oil–water mixtures.

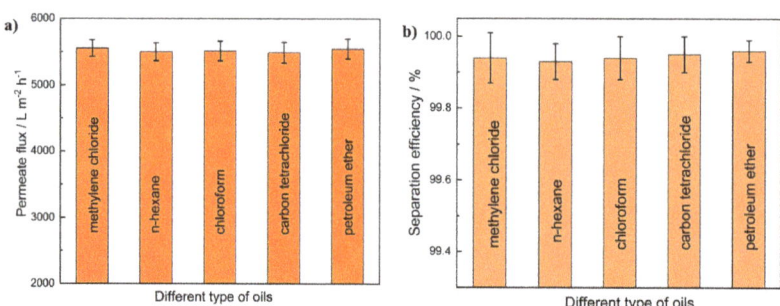

Figure 7. (a) Separation fluxes and (b) efficiencies of DP8–1.5 layered membrane for different oils.

3.5. Stability Test of PVDF/DP8/SiO$_2$ Composite Film

The long–term stability of the membrane is very important to the oil–water separation performance [54–57]. The membrane's water pollution resistance and self–cleaning ability were tested. In total, 4 µL deionized water was slowly and vertically dropped from the needle tube onto the DP8–1.5 composite membrane. When the water contacted the membrane surface in a large area, the water droplets were slowly lifted. Figure 8a clearly shows that the water droplets bounced off the surface, and the shape of the water droplets did not change during the entire experiment, did not fall off the needle under greater force, and did not adsorb to the surface, which shows that the modified film had a good anti–water adhesion and self–cleaning ability. As shown in Figure 8b, when the surface of the membrane was sprayed with water contaminated by methyl blue, the water rebounded and separated from the surface of the membrane, leaving no traces on the surface of the membrane. In addition, the methyl blue powder and sand grains were scattered on the surface of the membrane, and then the water was sprayed onto the surface of the membrane and rolled down the surface immediately. Moreover, the pollutants were taken away without leaving any traces and stains. In comparison, pure PVDF nanofiber membrane is easily contaminated under the same test [24,30].

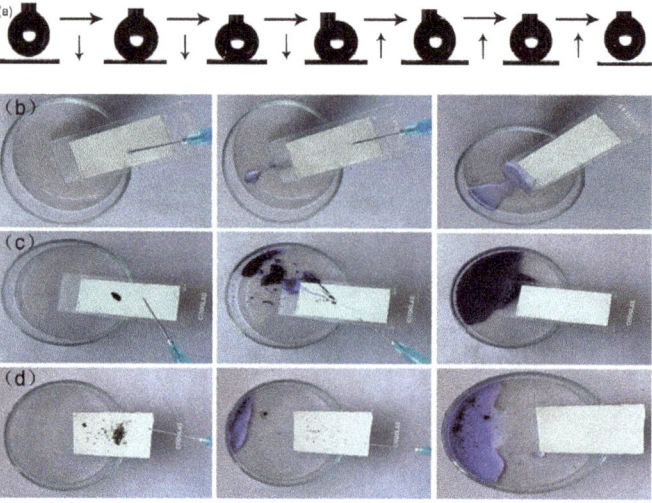

Figure 8. Self–cleaning ability of DP8–1.5 fiber membrane. (a) Water droplet drop experiment. (b) Water rebounds on the surface of an ultra–hydrophobic surface. (c) Methyl blue was removed by water droplets. (d) Sand grains were removed by water droplets.

To further study the stability of the membrane, the DP8–1.5 composite membrane was soaked in an acid–base salt solution for 24 h, and then the contact angle of the membrane was tested. Figure 9 shows the water contact angle (WCA) values immersed in the different solutions for different times. When immersed in an acid–base salt solution for 4, 8, 16, and 24 h, the contact angle of DP8–1.5 superhydrophobic composite film still showed superhydrophobicity. This shows that the DP8–1.5 layered structure membrane still showed excellent oil–water separation performance under different acid–base salt conditions.

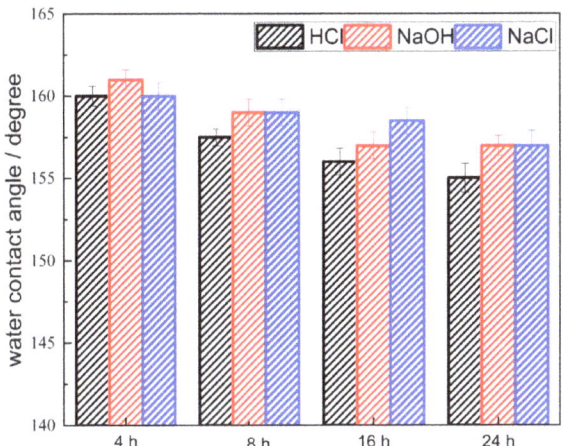

Figure 9. WCA test of DP8–1.5 layered membrane immersed in acid–base salt solution.

3.6. PVDF/DP8/SiO$_2$ Composite Film Surface Appearance and Mechanism Analysis

The structure of PVDF/DP8/SiO$_2$ was investigated by SEM. SEM images in Figure 10 showed the evolution of morphology of electrospun PVDF and electrospray DP8–X with different SiO$_2$ content. It is evident from a$_2$–g$_2$ of Figure 10 that the electrospun composite membrane had a distinct hierarchical structure. From Figure 10b$_1$–g$_1$, when SiO$_2$ and DP8 were added to the surface of the PVDF nanofiber matrix, microspheres with a hierarchical structure were obtained with the nanometer structure. It can be seen in Figure 10 b$_1$ that the surface of the microspheres formed by PVDF/DP8 is smooth. When SiO$_2$ was added, the surface of the microspheres was obviously wrinkled because SiO$_2$ was added to the microspheres. It can be seen from FTIR (Figure 1) that because DP8 itself ha phenolic hydroxyl groups, the phenolic hydroxyl groups themselves can be ionized. During electrospraying, the uncured microspheres formed an obvious honeycomb structure due to the interaction between the Coulomb force [41]. When the content of SiO$_2$ continued to increase to more than 2.5%, the surface honeycomb structure obviously disappeared (Figure 10b–e) which directly affected the performance of the membrane. As shown in WCA (Figure 6) test, with the increase in SiO$_2$ content, the oil flux increased significantly, mainly because the addition of SiO$_2$ can also form a good honeycomb structure, forming more oil channels. However, when the content of SiO$_2$ increased to 4%, the flux of the membrane was significantly reduced. When the higher content of SiO$_2$ was added, the hydrophobic gas phase SiO$_2$ nanoparticles coated the surface of the polymer and the phenolic hydroxyl functional groups of DP8 were covered. Hence, the uncured microspheres were not charged and could not generate Coulomb force, resulting in the disappearance of the honeycomb structure and reducing the oil channels and reducing the flux of oil–water separation during the electrospraying process.

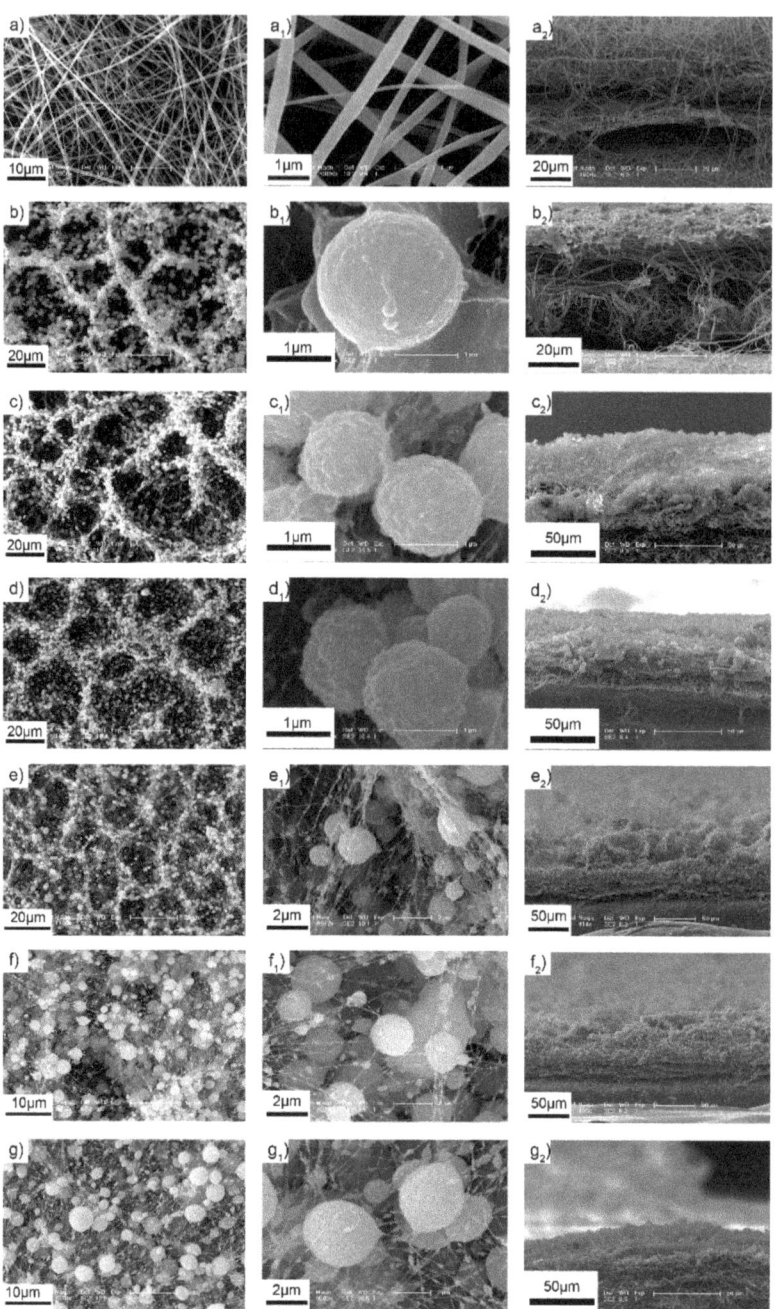

Figure 10. SEM of pure PVDF nanofiber membrane DP8 composite membrane and electrosprayed DP8–X with different SiO_2 content. (**a–a$_2$**) PVDF, (**b–b$_2$**) DP8, (**c–c$_2$**) DP8–0.5, (**d–d$_2$**) DP8–1.0, (**e–e$_2$**) DP8–1.5, (**f–f$_2$**) DP8–2.5, (**g–g$_2$**) DP8–4.0.

The hydrophobic model of the composite membrane is shown in Figure 11. PVDF was applied as the electrospinning substrate, and DP8 and SiO$_2$ were sprayed on the composite membrane by electrospraying. The Si–CH$_3$ bond in DP8 reduced its surface energy, and meanwhile, a honeycomb–like through–hole structure was formed due to the effect of electrostatic repulsion. This provided more channels and volume for conveying oil, thereby increasing oil flux. Moreover, the surface of the microspheres was coated with SiO$_2$ nanoparticles with a low surface energy, which increased the surface roughness of the membrane, and finally formed a composite membrane with a hierarchical structure and a honeycomb porous structure, which greatly enhanced the hydrophobicity.

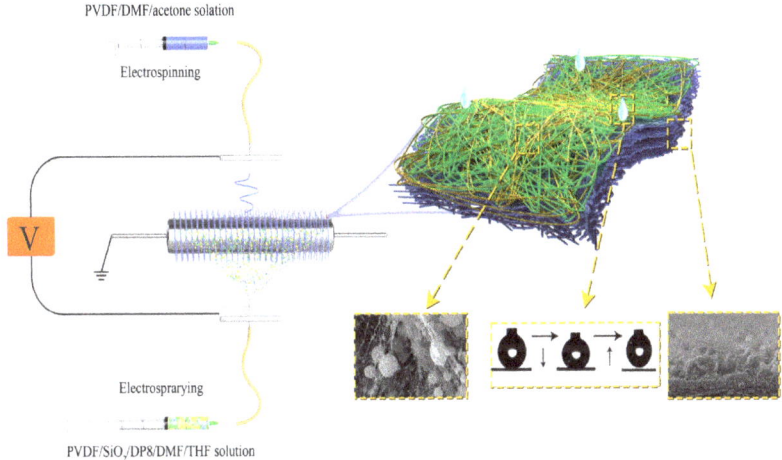

Figure 11. Hydrophobic model of PVDF/DP8/SiO$_2$ composite membrane.

3.7. Conclusions

This article mainly studies the preparation of PVDF/DP8/SiO$_2$ composite membranes with hierarchical structure and their performance in oil–water separation. During the electrospraying process, the addition of long segment polydimethylsiloxane and fumed SiO$_2$ formed pores through electrostatic repulsion between ions, thereby increasing the membrane flux, reducing its surface energy, and increasing its surface roughness. When the new polydimethylsiloxane DP8 and SiO$_2$ nanoparticles were added, the hydrophobicity of the film was greatly improved, and the hydrophobic WCA reached 162° ± 0.7°, which is much higher than pure PVDF. After 10 cycles of oil–water separation experiments, the separation efficiency of the composite membrane was still maintained above 99.94%. Because of the chemical composition, low surface energy, and honeycomb structure of the membrane, it has an ultra–high oil flux (>5400 L m^{-2} h^{-1}), separation efficiency, and self–cleaning ability.

Author Contributions: Conceptualization, C.Z. and Y.Y.; methodology, S.L.; software, Y.Y.; validation, B.L., C.Z. and C.C.; formal analysis, Y.Y.; investigation, B.L.; resources, S.W.; data curation, Y.Y.; writing—original draft preparation, Y.Y.; writing—review and editing, Y.Y.; visualization, C.Z.; supervision, C.Z.; project administration, C.C.; funding acquisition, S.W. All authors have read and agreed to the published version of the manuscript.

Funding: We thank the funding support from department of science and technology of Jilin Province (20220201111GX) and open research fund of state key laboratory of bio–based fiber manufacturing technology, China textile academy (SKL202203).

Institutional Review Board Statement: Not applicable.

Informed Consent Statement: Not applicable.

Data Availability Statement: Not applicable.

Conflicts of Interest: The authors declare no conflict of interest.

References

1. Gupta, R.K.; Dunderdale, G.J.; England, M.W.; Hozumi, A. Oil/Water Separation Techniques: A Review of Recent Progresses and Future Directions. *J. Mater. Chem. A* **2017**, *5*, 16025–16058. [CrossRef]
2. Yuan, T.; Meng, J.; Hao, T.; Wang, Z.; Zhang, Y. A Scalable Method toward Superhydrophilic and Underwater Superoleophobic PVDF Membranes for Effective Oil/Water Emulsion Separation. *ACS Appl. Mater. Interfaces* **2015**, *7*, 14896–14904. [CrossRef] [PubMed]
3. Mousa, H.M.; Fahmy, H.S.; Ali, G.A.M.; Abdelhamid, H.N.; Ateia, M. Membranes for Oil/Water Separation: A Review. *Adv. Mater. Interfaces* **2022**, *9*, 1–36. [CrossRef]
4. Ismail, N.H.; Salleh, W.N.W.; Ismail, A.F.; Hasbullah, H.; Yusof, N.; Aziz, F.; Jaafar, J. Hydrophilic Polymer–Based Membrane for Oily Wastewater Treatment: A Review. *Sep. Purif. Technol.* **2020**, *233*, 116007. [CrossRef]
5. Ge, J.; Zhang, J.; Wang, F.; Li, Z.; Yu, J.; Ding, B. Superhydrophilic and Underwater Superoleophobic Nanofibrous Membrane with Hierarchical Structured Skin for Effective Oil-in-Water Emulsion Separation. *J. Mater. Chem. A* **2017**, *5*, 497–502. [CrossRef]
6. Jernelv, A. The Threats from Oil Spills: Now, Then, and in the Future. *Ambio* **2010**, *39*, 353–366. [CrossRef] [PubMed]
7. Yue, X.; Li, Z.; Zhang, T.; Yang, D.; Qiu, F. Design and Fabrication of Superwetting Fiber-Based Membranes for Oil/Water Separation Applications. *Chem. Eng. J.* **2019**, *364*, 292–309. [CrossRef]
8. Guo, H.; Zhang, D.; Jiang, L. PAN/PVA Composite Nanofibrous Membranes for Separating Oil-in-Water Emulsion. *J. Polym. Res.* **2022**, *29*, 108. [CrossRef]
9. Zhu, Y.; Wang, D.; Jiang, L.; Jin, J. Recent Progress in Developing Advanced Membranes for Emulsified Oil/Water Separation. *NPG Asia Mater.* **2014**, *6*, e101. [CrossRef]
10. Teh, C.Y.; Budiman, P.M.; Shak, K.P.Y.; Wu, T.Y. Recent Advancement of Coagulation-Flocculation and Its Application in Wastewater Treatment. *Ind. Eng. Chem. Res.* **2016**, *55*, 4363–4389. [CrossRef]
11. Shen, B.; Du, C.; Wang, W.; Yu, D. Hydrophilic SPE/MPTES–PAN Electrospun Membrane Prepared via Click Chemistry for High Efficiency Oil–Water Separation. *J. Mater. Sci.* **2022**, *57*, 1474–1488. [CrossRef]
12. Jin, Y.; Huang, L.; Zheng, K.; Zhou, S. Blending Electrostatic Spinning Fabrication of Superhydrophilic/Underwater Superoleophobic Polysulfonamide/Polyvinylpyrrolidone Nanofibrous Membranes for Efficient Oil–Water Emulsion Separation. *Langmuir* **2022**, *38*, 8241–8251. [CrossRef] [PubMed]
13. Zhao, L.; Du, Z.; Tai, X.; Ma, Y. One-Step Facile Fabrication of Hydrophobic SiO_2 Coated Super–Hydrophobic/Super–Oleophilic Mesh via an Improved Stöber Method to Efficient Oil/Water Separation. *Colloids Surf. A Physicochem. Eng. Asp.* **2021**, *623*, 126404. [CrossRef]
14. Si, Y.; Guo, Z. Superwetting Materials of Oil–Water Emulsion Separation. *Chem. Lett.* **2015**, *44*, 874–883. [CrossRef]
15. Wang, Y.; Liu, Z.; Wei, X.; Liu, K.; Wang, J.; Hu, J.; Lin, J. An Integrated Strategy for Achieving Oil-in-Water Separation, Removal, and Anti–Oil/Dye/Bacteria-Fouling. *Chem. Eng. J.* **2021**, *413*, 127493. [CrossRef]
16. Chu, Z.; Feng, Y.; Seeger, S. Oil/Water Separation with Selective Superantiwetting/Superwetting Surface Materials. *Angew. Chem. Int. Ed.* **2015**, *54*, 2328–2338. [CrossRef]
17. Alotaibi, H.F.; Al Thaher, Y.; Perni, S.; Prokopovich, P. Role of Processing Parameters on Surface and Wetting Properties Controlling the Behaviour of Layer–by–Layer Coated Nanoparticles. *Curr. Opin. Colloid Interface Sci.* **2018**, *36*, 130–142. [CrossRef]
18. Kota, A.K.; Kwon, G.; Choi, W.; Mabry, J.M.; Tuteja, A. Hygro–Responsive Membranes for Effective Oilg-Water Separation. *Nat. Commun.* **2012**, *3*, 1025. [CrossRef]
19. Zhang, S.; Liu, H.; Tang, N.; Ge, J.; Yu, J.; Ding, B. Publisher Correction: Direct Electronetting of High–Performance Membranes Based on Self-Assembled 2D Nanoarchitectured Networks. *Nat. Commun.* **2020**, *11*, 41467. [CrossRef]
20. Nayak, K.; Tripathi, B.P. Molecularly Grafted PVDF Membranes with In–Air Superamphiphilicity and Underwater Superoleophobicity for Oil/Water Separation. *Sep. Purif. Technol.* **2021**, *259*, 118068. [CrossRef]
21. Zhou, Z.; Wu, X.F. Electrospinning Superhydrophobic–Superoleophilic Fibrous PVDF Membranes for High–Efficiency Water–Oil Separation. *Mater. Lett.* **2015**, *160*, 423–427. [CrossRef]
22. Ngang, H.P.; Ahmad, A.L.; Low, S.C.; Ooi, B.S. Preparation of Thermoresponsive $PVDF/SiO_2$–PNIPAM Mixed Matrix Membrane for Saline Oil Emulsion Separation and Its Cleaning Efficiency. *Desalination* **2017**, *408*, 1–12. [CrossRef]
23. Bae, J.; Kim, H.; Kim, K.S.; Choi, H. Effect of Asymmetric Wettability in Nanofiber Membrane by Electrospinning Technique on Separation of Oil/Water Emulsion. *Chemosphere* **2018**, *204*, 235–242. [CrossRef]
24. Yang, Y.; Li, Y.; Cao, L.; Wang, Y.; Li, L.; Li, W. Electrospun $PVDF–SiO_2$ Nanofibrous Membranes with Enhanced Surface Roughness for Oil–Water Coalescence Separation. *Sep. Purif. Technol.* **2021**, *269*, 118726. [CrossRef]
25. Du, C.; Wang, Z.; Liu, G.; Wang, W.; Yu, D. One–Step Electrospinning $PVDF/PVP–TiO_2$ Hydrophilic Nanofiber Membrane with Strong Oil–Water Separation and Anti–Fouling Property. *Colloids Surf. A Physicochem. Eng. Asp.* **2021**, *624*, 126790. [CrossRef]
26. Jadav, G.L.; Singh, P.S. Synthesis of Novel Silica–Polyamide Nanocomposite Membrane with Enhanced Properties. *J. Memb. Sci.* **2009**, *328*, 257–267. [CrossRef]

27. Myronchuk, V.G.; Dzyazko, Y.S.; Zmievskii, Y.G.; Ukrainets, A.I.; Bildukevich, A.V.; Kornienko, L.V.; Rozhdestvenskaya, L.M.; Palchik, A.V. Organic–Inorganic Membranes for Filtration of Corn Distillery. *Acta Period. Technol.* **2016**, *47*, 153–165. [CrossRef]
28. Pang, R.; Li, X.; Li, J.; Lu, Z.; Sun, X.; Wang, L. Preparation and Characterization of ZrO_2/PES Hybrid Ultrafiltration Membrane with Uniform ZrO_2 Nanoparticles. *Desalination* **2014**, *332*, 60–66. [CrossRef]
29. Zhang, Y.; Hu, Y.; Zhang, L.; Wang, Y.; Liu, W.; Ma, C.; Liu, S. Porous SiO_2 Coated $Al_xFe_yZr_{1-x-y}O_2$ Solid Superacid Nanoparticles with Negative Charge for Polyvinylidene Fluoride (PVDF) Membrane: Cleaning and Partial Desalinating Seawater. *J. Hazard. Mater.* **2020**, *384*, 121471. [CrossRef]
30. Gao, J.; Li, B.; Wang, L.; Huang, X.; Xue, H. Flexible Membranes with a Hierarchical Nanofiber/Microsphere Structure for Oil Adsorption and Oil/Water Separation. *J. Ind. Eng. Chem.* **2018**, *68*, 416–424. [CrossRef]
31. Chen, P.C.; Xu, Z.K. Mineral–Coated Polymer Membranes with Superhydrophilicity and Underwater Superoleophobicity for Effective Oil/Water Separation. *Sci. Rep.* **2013**, *3*, 2776. [CrossRef] [PubMed]
32. Zhang, F.; Zhang, W.B.; Shi, Z.; Wang, D.; Jin, J.; Jiang, L. Nanowire–Haired Inorganic Membranes with Superhydrophilicity and Underwater Ultralow Adhesive Superoleophobicity for High–Efficiency Oil/Water Separation. *Adv. Mater.* **2013**, *25*, 4192–4198. [CrossRef] [PubMed]
33. Zhang, W.; Shi, Z.; Zhang, F.; Liu, X.; Jin, J.; Jiang, L. Superhydrophobic and Superoleophilic PVDF Membranes for Effective Separation of Water–in–Oil Emulsions with High Flux. *Adv. Mater.* **2013**, *25*, 2071–2076. [CrossRef]
34. Zhang, Y.; Wu, L.; Wang, X.; Yu, J.; Ding, B. Super Hygroscopic Nanofibrous Membrane–Based Moisture Pump for Solar–Driven Indoor Dehumidification. *Nat. Commun.* **2020**, *11*, 3302. [CrossRef]
35. Tropmann, A.; Tanguy, L.; Koltay, P.; Zengerle, R.; Riegger, L. Completely Superhydrophobic PDMS Surfaces for Microfluidics. *Langmuir* **2012**, *28*, 8292–8295. [CrossRef] [PubMed]
36. Park, E.J.; Cho, Y.K.; Kim, D.H.; Jeong, M.G.; Kim, Y.H.; Kim, Y.D. Hydrophobic Polydimethylsiloxane (PDMS) Coating of Mesoporous Silica and Its Use as a Preconcentrating Agent of Gas Analytes. *Langmuir* **2014**, *30*, 10256–10262. [CrossRef]
37. Ma, W.; Zhang, Q.; Hua, D.; Xiong, R.; Zhao, J.; Rao, W.; Huang, S.; Zhan, X.; Chen, F.; Huang, C. Electrospun Fibers for Oil–Water Separation. *RSC Adv.* **2016**, *6*, 12868–12884. [CrossRef]
38. Lee, E.J.; Deka, B.J.; Guo, J.; Woo, Y.C.; Shon, H.K.; An, A.K. Engineering the Re–Entrant Hierarchy and Surface Energy of PDMS–PVDF Membrane for Membrane Distillation Using a Facile and Benign Microsphere Coating. *Environ. Sci. Technol.* **2017**, *51*, 10117–10126. [CrossRef]
39. Jia, W.; Kharraz, J.A.; Choi, P.J.; Guo, J.; Deka, B.J.; An, A.K. Superhydrophobic Membrane by Hierarchically Structured PDMS–POSS Electrospray Coating with Cauliflower–Shaped Beads for Enhanced MD Performance. *J. Memb. Sci.* **2020**, *597*, 117638. [CrossRef]
40. Li, H.; Zhao, X.; Wu, P.; Zhang, S.; Geng, B. Facile Preparation of Superhydrophobic and Superoleophilic Porous Polymer Membranes for Oil/Water Separation from a Polyarylester Polydimethylsiloxane Block Copolymer. *J. Mater. Sci.* **2016**, *51*, 3211–3218. [CrossRef]
41. Luo, S.; Dai, X.; Sui, Y.; Li, P.; Zhang, C. Preparation of Biomimetic Membrane with Hierarchical Structure and Honeycombed Through–Hole for Enhanced Oil–Water Separation Performance. *Polymer* **2021**, *218*, 123522. [CrossRef]
42. Wu, J.; Ding, Y.; Wang, J.; Li, T.; Lin, H.; Wang, J.; Liu, F. Facile Fabrication of Nanofiber- and Micro/Nanosphere-Coordinated PVDF Membrane with Ultrahigh Permeability of Viscous Water-in-Oil Emulsions. *J. Mater. Chem. A* **2018**, *6*, 7014–7020. [CrossRef]
43. Gao, J.; Huang, X.; Xue, H.; Tang, L.; Li, R.K.Y. Facile Preparation of Hybrid Microspheres for Super–Hydrophobic Coating and Oil–Water Separation. *Chem. Eng. J.* **2017**, *326*, 443–453. [CrossRef]
44. Yuan, X.T.; Xu, C.X.; Geng, H.Z.; Ji, Q.; Wang, L.; He, B.; Jiang, Y.; Kong, J.; Li, J. Multifunctional PVDF/CNT/GO Mixed Matrix Membranes for Ultrafiltration and Fouling Detection. *J. Hazard. Mater.* **2020**, *384*, 120978. [CrossRef]
45. Meringolo, C.; Mastropietro, T.F.; Poerio, T.; Fontananova, E.; De Filpo, G.; Curcio, E.; Di Profio, G. Tailoring PVDF Membranes Surface Topography and Hydrophobicity by a Sustainable Two–Steps Phase Separation Process. *ACS Sustain. Chem. Eng.* **2018**, *6*, 10069–10077. [CrossRef]
46. Abdel–Hakim, A.; El–Basheer, T.M.; Abdelkhalik, A. Mechanical, Acoustical and Flammability Properties of SBR and SBR–PU Foam Layered Structure. *Polym. Test.* **2020**, *88*, 106536. [CrossRef]
47. Abdel–Hakim, A.; El–Mogy, S.A.; EL–Zayat, M.M. Radiation Crosslinking of Acrylic Rubber/Styrene Butadiene Rubber Blends Containing Polyfunctional Monomers. *Radiat. Phys. Chem.* **2019**, *157*, 91–96. [CrossRef]
48. Abdelkhalik, A.; Abdel–Hakim, A.; Makhlouf, G.; El–Gamal, A.A. Effect of Iron Poly(Acrylic Acid–Co–Acrylamide) and Melamine Polyphosphate on the Flammability Properties of Linear Low–Density Polyethylene. *J. Therm. Anal. Calorim.* **2019**, *138*, 1021–1031. [CrossRef]
49. Chen, L.; Si, Y.; Zhu, H.; Jiang, T.; Guo, Z. A Study on the Fabrication of Porous PVDF Membranes by In-Situ Elimination and Their Applications in Separating Oil/Water Mixtures and Nano-Emulsions. *J. Memb. Sci.* **2016**, *520*, 760–768. [CrossRef]
50. Li, J.; Guo, S.; Xu, Z.; Li, J.; Pan, Z.; Du, Z.; Cheng, F. Preparation of Omniphobic PVDF Membranes with Silica Nanoparticles for Treating Coking Wastewater Using Direct Contact Membrane Distillation: Electrostatic Adsorption vs. Chemical Bonding. *J. Memb. Sci.* **2019**, *574*, 349–357. [CrossRef]
51. Zhang, R.; Xu, Y.; Shen, L.; Li, R.; Lin, H. Preparation of Nickel@polyvinyl Alcohol (PVA) Conductive Membranes to Couple a Novel Electrocoagulation–Membrane Separation System for Efficient Oil-Water Separation. *J. Memb. Sci.* **2022**, *653*, 120541. [CrossRef]

52. Zhang, Z.M.; Gan, Z.Q.; Bao, R.Y.; Ke, K.; Liu, Z.Y.; Yang, M.B.; Yang, W. Green and Robust Superhydrophilic Electrospun Stereocomplex Polylactide Membranes: Multifunctional Oil/Water Separation and Self–Cleaning. *J. Memb. Sci.* **2020**, *593*, 117420. [CrossRef]
53. Jabbarnia, A.; Khan, W.S.; Ghazinezami, A.; Asmatulu, R. Investigating the Thermal, Mechanical, and Electrochemical Properties of PVdF/PVP Nanofibrous Membranes for Supercapacitor Applications. *J. Appl. Polym. Sci.* **2016**, *133*, 43707. [CrossRef]
54. Zhu, Y.; Xie, W.; Zhang, F.; Xing, T.; Jin, J. Superhydrophilic In–Situ–Cross–Linked Zwitterionic Polyelectrolyte/PVDF–Blend Membrane for Highly Efficient Oil/Water Emulsion Separation. *ACS Appl. Mater. Interfaces* **2017**, *9*, 9603–9613. [CrossRef]
55. Tao, M.; Xue, L.; Liu, F.; Jiang, L. An Intelligent Superwetting PVDF Membrane Showing Switchable Transport Performance for Oil/Water Separation. *Adv. Mater.* **2014**, *26*, 2943–2948. [CrossRef]
56. Zhang, W.; Zhu, Y.; Liu, X.; Wang, D.; Li, J.; Jiang, L.; Jin, J. Salt–Induced Fabrication of Superhydrophilic and Underwater Superoleophobic PAA–g–PVDF Membranes for Effective Separation of Oil–in–Water Emulsions. *Angew. Chem. Int. Ed.* **2014**, *53*, 856–860. [CrossRef]
57. Xiang, Y.; Shen, J.; Wang, Y.; Liu, F.; Xue, L. A PH–Responsive PVDF Membrane with Superwetting Properties for the Separation of Oil and Water. *RSC Adv.* **2015**, *5*, 23530–23539. [CrossRef]

Article

Controlling Morphology and Wettability of Intrinsically Superhydrophobic Copper-Based Surfaces by Electrodeposition

Raziyeh Akbari [1,2,3,*], Mohammad Reza Mohammadizadeh [1], Carlo Antonini [3], Frédéric Guittard [2] and Thierry Darmanin [2]

[1] Supermaterials Research Laboratory (SRL), Department of Physics, University of Tehran, North Kargar Ave., Tehran P.O. Box 14395-547, Iran
[2] NICE Lab, Université Côte d'Azur, Parc Valrose, 06108 Nice, France
[3] Department of Materials Science, University of Milano-Bicocca, Via R. Cozzi 55, 20125 Milano, Italy
* Correspondence: raziyeh.akbari@unimib.it

Citation: Akbari, R.; Mohammadizadeh, M.R.; Antonini, C.; Guittard, F.; Darmanin, T. Controlling Morphology and Wettability of Intrinsically Superhydrophobic Copper-Based Surfaces by Electrodeposition. *Coatings* 2022, 12, 1260. https://doi.org/10.3390/coatings12091260

Academic Editor: Paweł Nowak

Received: 27 July 2022
Accepted: 25 August 2022
Published: 29 August 2022

Publisher's Note: MDPI stays neutral with regard to jurisdictional claims in published maps and institutional affiliations.

Copyright: © 2022 by the authors. Licensee MDPI, Basel, Switzerland. This article is an open access article distributed under the terms and conditions of the Creative Commons Attribution (CC BY) license (https://creativecommons.org/licenses/by/4.0/).

Abstract: Electrodeposition is an effective and scalable method to grow desired structures on solid surfaces, for example, to impart superhydrophobicity. Specifically, copper microcrystals can be grown using electrodeposition by controlling deposition parameters such as the electrolyte and its acidity, the bath temperature, and the potential modulation. The aim of the present work is the fabrication of superhydrophobic copper-based surfaces by electrodeposition, investigating both surface properties and assessing durability under conditions relevant to real applications. Accordingly, copper-based layers were fabricated on Au/Si(100) from $Cu(BF_4)_2$ precursor by electrodeposition, using cyclic voltammetry and square-pulse voltage approaches. By increasing the bath temperature from 22 °C to 60 °C, the growth of various structures, including micrometric polyhedral crystals and hierarchical structures, ranging from small grains to pine-needle-like dendrite leaves, has been demonstrated. Without any further physical and/or chemical modification, samples fabricated with square-pulse voltage at 60 °C are superhydrophobic, with a contact angle of 160° and a sliding angle of 15°. In addition, samples fabricated from fluoroborate precursor are carefully compared to those fabricated from sulphate precursor to compare chemical composition, surface morphology, wetting properties, and durability under UV exposure and hard abrasion. Results show that although electrodeposition from fluoroborate precursor can provide dendritic microstructures with good superhydrophobicity properties, surfaces possess lower durability and stability compared to those fabricated from the sulphate precursor. Hence, from an application point of view, fabrication of copper superhydrophobic surfaces from sulphate precursor is more recommended.

Keywords: wetting; electrodeposition; nanostructure; copper; durability

1. Introduction

In recent years, the use of water-repellent surfaces has increased in industrial applications due to their application as self-cleaning, antibacterial and anticorrosion surfaces, as well as drag reduction in marine applications [1–4]. Water repellence is achieved by minimizing the contact area between the surface and the water drop, and can be understood by studying the static contact angle θ_S [5]. As argued by Cassie and Baxter, surface hydrophobicity is correlated to the surface texture and to the material's intrinsic hydrophobicity [6–8], which is determined by the chemical composition. According to the Cassie-Baxter model, air pockets trapped inside the surface textures can prevent liquid from penetrating grooves, increasing water repellency and reducing drop–substrate adhesion. Specifically, complex hierarchical structures with micro- and nanometer features have been found to be ideal for stable superhydrophobicity [6,9,10]. To assess surface wetting properties in addition to static contact angle, θ_S, a study on the wetting hysteresis in quasi-static conditions is also crucial for investigating the wetting state of a solid surface. The wetting hysteresis (H) is defined as the difference between the largest and lowest contact angles of the drop on the solid surface, i.e., the advancing (θ_{Adv}) and the receding (θ_{Rec}) contact angles, respectively (being $H = \theta_{Adv} - \theta_{Rec}$). In the definition of a superhydrophobic sample from an application view, the sliding angle (SA), which is defined as the minimum tilting angle of surface where the water drop easily slides on the surface, is also important. According to literatures [5,6,9,10], H and SA should ideally be lower than 10°.

Among the different surface-fabrication methods, electrodeposition is an effective and scalable method for the fabrication of textured surfaces with micrometric crystals of conductive materials, including metals, for example, copper for water-repellent surfaces [11,12] and cobalt–nickel alloy for anticorrosion application [13] and conductive polymers for porous structures [14]. Furthermore, electrodeposition is a cost-effective, scalable, and relatively fast method, which is already widely used in several industrial sectors for surface treatment and finishing. Indeed, electrodeposition can be used to control the surface roughness and morphology for the discovery and development of novel nanostructured materials with good mechanical properties [15,16]. Surface structures, spanning from 1D needles and fibres to 2D ribbons and sheets, and 3D hollow spheres, dendrites, and flower-like structures, can be used in many applications such as microelectronics, optoelectronics, lithium batteries, and biomedical applications [16–18]. In electrochemical deposition, also referred to as electrodeposition and electroplating, the electrolyte cell contains a working electrode (i.e., substrate), a counter electrode, and a reference electrode, immersed in an ionic conductor electrolyte solution (see schematic in Figure 1). Metal electrodeposition on the substrate occurs through the electrochemical reduction of ions from the electrolyte by applying an electric potential between the cathode (i.e., working electrode) and the anode (i.e., counter electrode). The potential drives the positive ions to migrate toward the extra electrons near the negatively charged cathode. During the deposition, metal ions are reduced and form a crystalline structure on the substate surface. The layer thickness is mainly determined by the electrodeposition conditions, including electrolyte chemical composition, bath temperature, deposition time, current, voltage, and modulation method [17,19–22]. Increasing the electrodeposition time in a highly reactive environment, for example, by increasing the number of deposition cycles in the cyclic voltammetry, or increasing the number of pulses in the pulse-voltage electrodeposition, results in a more intense and rapid deposition, which can increase the growth of more complex and vertical structures on the surface [17]. According to studies relating wettability to surface roughness [6,9,10,23], surface morphology, combined with the surface chemistry, has a significant influence on the wettability. Briefly, an increase in surface roughness enhances the surface properties, and thus can make a hydrophobic surface even more hydrophobic. Higher roughness can be achieved by increasing the deposition time, and eventually reaching superhydrophobicity.

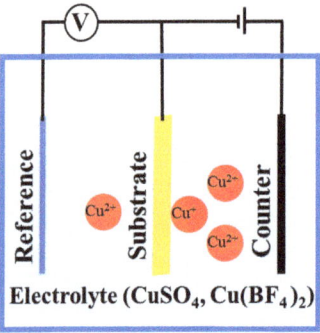

Figure 1. A schematic of an electrodeposition cell for depositing copper from an electrolyte containing copper sulphate or copper tetrafluoroborate solutions. A counter electrode repulses copper ions to the substrate. A reference electrode controls the reaction.

Since copper is widely used in several industrial networks, including water and electricity networks, solar energy and transportation [24–26], adding superhydrophobicity to the copper-based industrial surfaces will apply self-cleaning capability to these surfaces and increase the durability of the surfaces under wet conditions. To create copper-based layers using electrodeposition, various precursors are used in the industry, including sulphates, fluoroborates, acetates, alkyls and chlorides [17,27–30]. Previous studies have shown that the required potential for copper electrodeposition from the acid solution is less than that from alkaline solution, due to the higher conductivity and lower electrode polarization [17,30]. Hence, electrode polarization can be neglected for low current densities in acidic solutions, where the deposition rate is relatively higher. For the specific target of the fabrication of superhydrophobic copper-based surfaces, evidence is needed to identify which copper precursor provides better performances in controlling surface wetting properties. Among all, fluoroborate and sulphate precursors have the highest application potential for a variety of reasons: fluoroborate can lead to thicker layers; sulphate leads to homogeneously shaped layers and is abundant in the mineral residues. Thus, since the acidity of copper fluoroborate is higher than other acidic copper precursors such as sulphates, it does not require addition of acids to increase the electrochemical activity of the electrolyte [17]. By using copper fluoroborate solution dissolved in dodecylbenzene sulphonic acid sodium salt (DBSA) and poly(vinylpyrrolidone) (PVP), which act as ion stabilizer in the electrolyte solution, Ko et al. [31] reported the fabrication of various copper architectures such as pyramids, cubes, and multipods. Tetrahedral pyramids were grown in the ratio of 1:3 from copper and DBSA solution [31]. By decreasing the ratio to 1:2, the created crystals on the surface converted to free-standing cubes [32], whereas multipods were observed for copper solution in PVP [33]. PVP acts as a capping surfactant reagent, which is adsorbed differently on various crystal surfaces and leads to a competitive growth between different copper crystal facets, and results in shape variation of the final crystal shape. According to the literature [31,34–37], some of possible reactions in an aqueous solution of copper during the electrodeposition are as follows:

$$Cu^{2+} + 2e^- \rightarrow Cu\ (s) \tag{1}$$

$$Cu^{2+} + e^- \rightarrow Cu^+ \tag{2}$$

$$Cu^+ + e^- \rightarrow Cu\ (s) \tag{3}$$

$$2Cu^+ + 2OH^- \rightarrow Cu_2O\ (s) + H_2O \tag{4}$$

$$Cu^{2+} + 2OH^- \rightarrow CuO\ (s) + H_2O \tag{5}$$

The application of a negative voltage between the counter electrode and the substrate can convert Cu^{2+} ions to Cu deposits on the substrate (Equation (1)) if the absolute value of voltages is higher than the reduction potential. For an absolute value of voltages lower than the reduction potential, there is also a possibility of a two-step reduction in the Cu^{2+} ions to Cu: firstly, Cu^{2+} converts to Cu^+, and then Cu^+ absorbs an electron and reduces to Cu deposits on the substrate, following Equations (2) and (3), respectively. A third reduction mechanism is based on partial reaction of copper ions with hydroxide ions in the electrolyte following Equations (4) and (5). These conditions can lead to an increase in the copper oxide content of the deposited layer in the higher temperatures. In addition, the reactivity and wetting of the as-prepared copper layers from aqueous solution can also increase the amount of adsorbed oxygen on the layer surface, which can be higher in highly rough-structured layers [28,29,38].

The present study first conducts a systematic investigation of the electrodeposition of fluoroborate precursors, using both cyclic voltammetry and square-pulse voltage in an aqueous solution. All samples from fluoroborate precursors (referred to as fluoroborate samples or fluoroborate surfaces for brevity) are characterized by goniometry, profilometry, scanning electron microscopy, and X-ray diffraction, to correlate electrodeposition parameters to surface morphology and wettability. The aim of the present work is the fabrication of superhydrophobic copper-based samples and investigating its surface properties and durability for real applications. Thus, following our previous studies on samples from copper sulphate precursor [27–29,38] (referred to as sulphate samples or sulphate surfaces for brevity), a comparison between the use of the two precursors is presented here to provide an overall assessment of surface morphology, chemistry, hydrophobicity, and durability.

2. Materials and Methods

2.1. Electrodeposition Conditions

An aqueous solution from 0.1 M copper (tetra-)fluoroborate precursor (Sigma Aldrich, St. Louis, MI, USA) with pH = 3.15 was prepared to deposit copper and copper oxides on Au/Si(100) substrates using both cyclic voltammetry and square-pulse voltage at bath temperatures of 22, 45, and 60 °C (see [27,29,38]). The electrochemical system included an Autolab potentiostat of Metrohm with three connected electrodes: (i) a 150 nm Au on Si(100) wafer as working electrode, (ii) a carbon rod as counter electrode, and (iii) a saturated calomel (SCE) as reference electrode. In the squared-pulse voltage deposition (referred to as "pulse" for convenience), each deposition cycle consisted of 10 s deposition at a fixed working voltage of $E_W = -0.3$ V and subsequently 2 s relaxing at 0 V, following our previous studies [27–29,38]. Deposition cycles were repeated 8 or 12 times at three bath temperatures. In cyclic voltammetry deposition (referred to "CV"), cycles were repeated 3 or 5 times at three bath temperatures, with voltage in the range $[-0.3, 0]$ V and a scan rate of 20 mV/s. The prepared samples were washed in distilled water and dried for one week in a sealed glass box in ambient conditions before characterization.

2.2. Surface Characterization

A Wyko NT1100 optical microscope (Bruker) with high-magnification vertical scanning interferometry (VSI), a field of view 0.5X, objective 50X, and scan size 239 × 182 µm was used to measure the average surface roughness ($R_{a,O}$). $R_{a,O}$ is measured noncontact using optical interferences. Since there is a divergence between roughness numbers measured by stylus and optical methods (see [39–41]), specially in highly rough and dendrite surfaces, the roughness measured in this article is named as $R_{a,O}$ to indicate the measurement method. Surface morphology was imaged by scanning electron microscopy (SEM) machines including 6700F JEOL and Vega TS5136 XM Tescan microscopes. A Philips XRD X'pert MPD diffractometer (Cu Kα radiation, 1.54 Å) with a step size of 0.02° and count time of 1 s per step in 2θ, ranging from 10° to 80°, was used to provide X-ray diffraction (XRD) patterns of the samples. The wetting properties of the samples were measured using a DSA30 goniometer (Krüss) as well as an in-home contact angle measurement set-up consisting of a high-speed camera (PHOTRON-NOVA FASTCAM S6, 1:1 Tokina AT-X M100 PRO D lens, 20 µm pixel size) with 2 to 7 µL water drops (γLV = 72.8 mN/m), repeating the measurement in at least three different positions for each sample. The drop size for sliding-angle (SA) measurements was 5 µL. The contact angle images and videos were analyzed using Dropen, an open-source in-house-developed software [42]. A low-intensity UV oven (SHAREBOT UCB), 405 nm wavelength, 120 W power, with ~20 cm lamp–sample distance, was used for surface cleaning in periods of 5 to 150 minutes. Abrasion tests were performed using a dedicated test setup consisting of sandpaper, a weight, and a ruler. As shown in the side view in Figure 2 the back side of the sample was attached to a glass slide using an adhesive tape and placed on P1500 SiC sandpaper. A 100 g weight was placed on the glass to increase the contact and pressure between the sample and the sandpaper. The test was performed by pulling the sample on the sandpaper along different distances, up to 30 cm. After every 3 cm of abrasion, the static and quasi-static (i.e., advancing and receding) contact angles were measured.

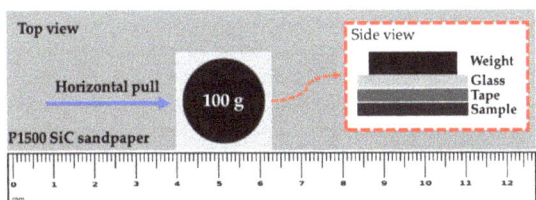

Figure 2. A schematic of the abrasion test setup used in this paper. It consisted of sandpaper, weight, and ruler. As shown in the side view, the sample was attached to a glass slide from its backside and placed on the sandpaper. The test was performed by pulling the sample on the sandpaper along different distances, as a 100 g weight was placed on the sample. After every 3 cm, the wetting state of the sample was examined.

3. Results

The results of the electrodeposition of copper from 0.1 M copper (tetra-)fluoroborate at 22, 45, and 60 °C using square potential pulse and cyclic voltammetry methods are presented, including XRD spectrum, contact angle values, roughness, and SEM images to visualize surface morphology.

3.1. Square Pulse

Chemistry and morphology of the prepared samples using the square-pulse method have been investigated to find the influence of the surface characteristics on the hydrophobicity. The XRD measurements show that Cu and Cu_2O facets in (111) direction are the only components in the deposited layer (see Figure 3). According to the results, the amount of copper is decreased by increasing the bath temperature while the amount of copper oxide is increasing, whereas at 60 °C, copper and copper oxide have similar intensities. A similar behaviour is observed in the sulphate samples [38]. Previous studies [23,27,38] have shown that due to high reactivity of copper structures especially in noncrystalline, highly rough, and fine-grained wet surfaces, it is expected that highly rough copper surfaces partially react with air under the ambient conditions after deposition while they are still wet and reactive. Hence, a slight change in the surface chemical composition towards more oxidation is considerable in this condition.

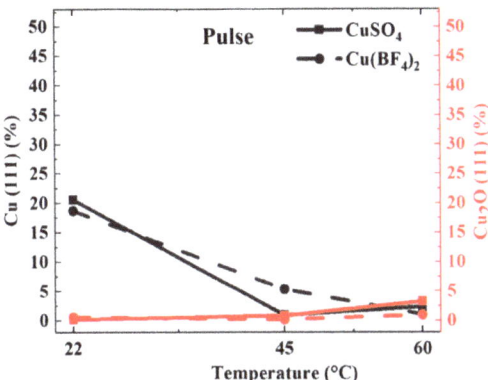

Figure 3. A comparison between the main Cu and Cu_2O peaks (both in (111) direction) in XRD spectrum of the prepared samples by pulse method using sulphate and fluoroborate precursors. The relative intensities are reported in % in comparison to the largest XRD peak of the substrate (i.e., Au (111)).

Figure 4a,b show that with every 15 °C increase in the deposition temperature, surface roughness increases by an order of magnitude. Hence, the roughness at 22, 45, and 60 °C is around 20 nm, 200 nm, and 1 µm, respectively. In addition, wetting measurements (see Figure 4a,b) show that θ_S increases by increasing the surface roughness as well as the number of pulses. In addition, the highest θ_S is observed in samples of 60 °C, reaching 156° in 8 pulses, and increasing to 160° in 12 pulses with five times larger roughness. This large roughness and superhydrophobic behaviour indicate the formation of a complex hierarchical structure, which causes the improvement of the surface water repellency, according to the Cassie–Baxter equation [6]. This is confirmed by SEM images, Figure 5.

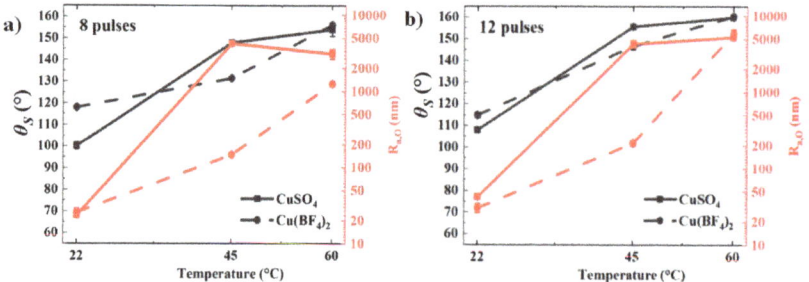

Figure 4. A comparison in contact angle and roughness of prepared samples from sulphate and fluoroborate solutions using pulse method in (a) 8 and (b) 12 numbers of deposition pulses. θ_S of the uncoated substrate is 85°.

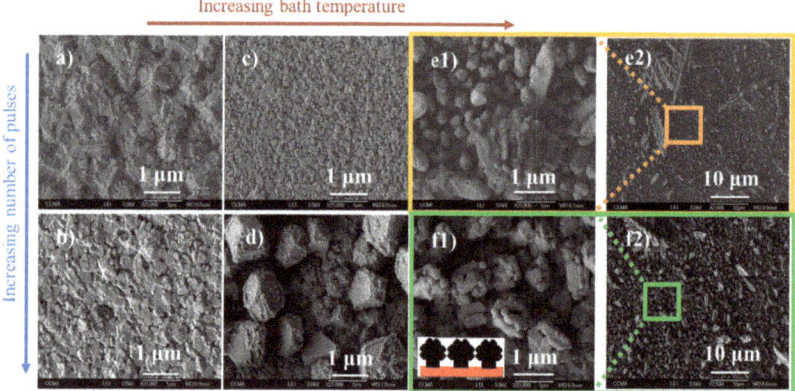

Figure 5. SEM images of the samples prepared from fluoroborate precursor using pulse method. (a) 22 °C, 8 pulses, nominal grain size (D_g) ≈ 300–500 nm; (b) 22 °C, 12 pulses, D_g ≈ 200–400 nm; (c) 45 °C, 8 pulses, D_g ≈ 60–200 nm; (d) 45 °C, 12 pulses, D_g ≈ 1–1.3 µm; (e1) 60 °C, 8 pulses, D_g ≈ 500 nm–1.2 µm; (e2) a larger view of e1 with fractal trees of around 15 µm; (f1) 60 °C, 12 pulses, D_g ≈ 300–600 nm, presented in a larger view in (f2) with fractal trees with different lengths from 3 to 20 µm. A schematic pattern of hierarchical structures is shown in f1.

According to Figure 5a, the structure formed at 22 °C is an accumulation of crystallites, a few hundred nm in diameter, with flat facets in various tetrahedron shapes. The size and out-of-plane growth of the structures are decreased by increasing the number of pulses (Figure 5b). In addition, although most of the crystal facets are tetrahedrons in 8 pulses, they are more likely to be broken triangles (i.e., hexahedron) in 12 pulses. Thus, the ability to follow the growth regime is weakened by increasing the number of deposited layers. At 45 °C, initially, in 8 pulses, a similar structure to 22 °C is grown (Figure 5c). By increasing the deposition pulses (Figure 5d), the size of crystals is increased significantly contradictory to the deposition in ambient temperature. This has led to the formation of micrometric octahedral crystals with deep valleys in between and increases θ_S to 147°. At

60 °C, changes in the surface structure are more prominent where different dendrite and hierarchical structures are formed on the sample (see Figure 5e1,e2), which is a result of instabilities in the solvent at this temperature due to approaching the boiling temperature of the solution. According to our experiments, the deposition rate in fluoroborate solution is considerably low in bath temperatures higher than 60 °C. The formed structures include (1) standing fractal leaves in 15μm length, longer than our previous samples prepared by copper sulphate precursor [28,38]; (2) Step-like hierarchical structures formed upon a larger crystal with an average diameter of 1 μm; (3) 3D octahedral pyramids with diameters ranging from 300 to 500 nm, similar to [29]. The formation of this complex structure leads to a sharp increase in surface roughness, resulting in an increase in θ_S to 155° for 8 pulses. However, the drop remains stuck on the surface and the sample is not superhydrophobic, an intermediate state between Wenzel and Cassie–Baxter [9], possibly due to the heterogeneity of surface roughness. By increasing the deposition to 12 pulses, partially ordered pine-like structures with lengths of 3 to 20 μm are observed on the surface (see Figure 5f2). Therefore, increasing the number of sharp vertical needles and more ordering and hierarchy in the small crystals formed between large structures, as well as critically larger roughness, lead to an increase in θ_S to 160°, where the water drop rolls down from the surface by tilting the surface to 15° for a 5 μL drop.

3.2. Cyclic Voltammetry

To study the changes in the surface hydrophobicity by electrodepositing copper from the fluoroborate solution, samples prepared by cyclic voltammetry are investigated through the chemical and physical characteristics. According to Figure 6, the cyclic voltammetry samples are covered by Cu(111). The amount of Cu_2O in the samples is negligible. According to the last observations [27–29,38], the very high intensity for Cu at 60 °C in respect to the lower temperature could be a result of the formation of a significantly thicker layer.

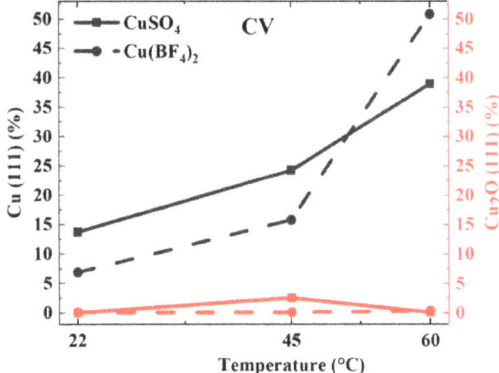

Figure 6. A comparison between the main Cu and Cu_2O peaks (both in (111) direction) in XRD spectrum of the prepared samples by CV method using sulphate and fluoroborate precursors. The relative intensities are reported in % in comparison to the largest XRD peak of the substrate (i.e., Au (111)).

Figure 7a,b show that the roughness is increased significantly by increasing the bath temperature and the number of deposition cycles. In '60 °C, 5 cycles', the roughness reaches the maximum of ~2μm due to creation of complex hierarchical structures. According to the θ_S results, the hydrophobicity in the samples follows the roughness changes and increases by increasing the bath temperature and the number of deposition cycles. Hence, the wetting in CV samples is in the Wenzel state and reaches 137° at maximum roughness.

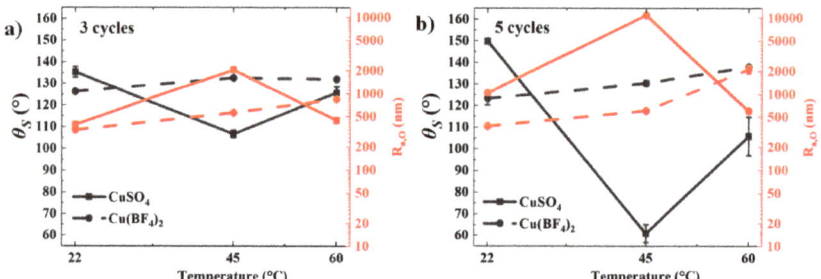

Figure 7. A comparison in contact angle and roughness of prepared samples from sulphate and fluoroborate solutions using CV method in (**a**) 3 and (**b**) 5 numbers of deposition cycles. θ_S of the uncoated substrate is 85°.

The surface of the '22 °C, 3 cycles' sample shown in Figure 8a consists of submicron octahedral crystallites with flat facets formed between smaller flat crystals, 1μm in diameter. The size of the small structures formed between large structures is increased by further deposition at the same bath temperature, i.e., '22 °C, 5 cycles', whereas the number and the size of the upper flat crystals are reduced (see Figure 8b). Thus, the growth of large structures and their out-of-plane growth were intensively increased by increasing the number of deposition cycles. Crystallites twice the size are grown at '45 °C, 3 cycles' compared to the samples fabricated in 22 °C (Figure 8c) while the shape of the crystallites is almost similar in these two bath temperatures. The size of structures and their out-of-plane growth are increased significantly by increasing the number of deposition cycles in 45 °C, whereas larger crystals ~5 μm in diameter and deeper and wider valleys are formed on the surfaces fabricated in 5 cycles (Figure 8d). The growth regime of the surface structure, i.e., the growth of micrometric surface crystallites with flat facets, as well as an increase in the size and the out-of-plane growth of crystallites by increasing the bath temperature and the number of deposition cycles, are followed at '60 °C, 3 cycles' (Figure 8e1,e2), whereas the size of the crystals is ~4 times larger than '45 °C, 3 cycles'. Thus, increasing the bath temperature prepares good conditions for the growth of large microcrystals and increases the crystallinity of the sample, in agreement with the previous observations [43–46]. Moreover, due to the increase in the mobility of copper ions towards the substrate at 60 °C, a temperature close to the boiling point of the electrolyte solution, some vertical aggregated structures are also observed in a larger view to the surface of '60 °C, 3 cycles', Figure 8e2. By increasing the number of deposition cycles at 60 °C, the vertical growth was expanded and became the prominent surface growth regime. Thus, vertical leaves with a length of ~3–10 μm are effectively grown in '60 °C, 5 cycles' (Figure 8f1,f2). This dendritic growth is due to the increase in the formation of air bubbles at 60 °C, close to the boiling point of the solution, and subsequent instability in the electrochemical conditions of the deposition. The surface between the leaves is covered with an aggregation of broken flat crystallites (Figure 8f1), which is smaller than the pulse samples (see Figure 5f1). The shape of the surface structures in '60 °C, 5 cycles', including the dendrite vertical leaves with sharp tips and the broken pyramids between the leaves, is substituted into Figure 8f2.

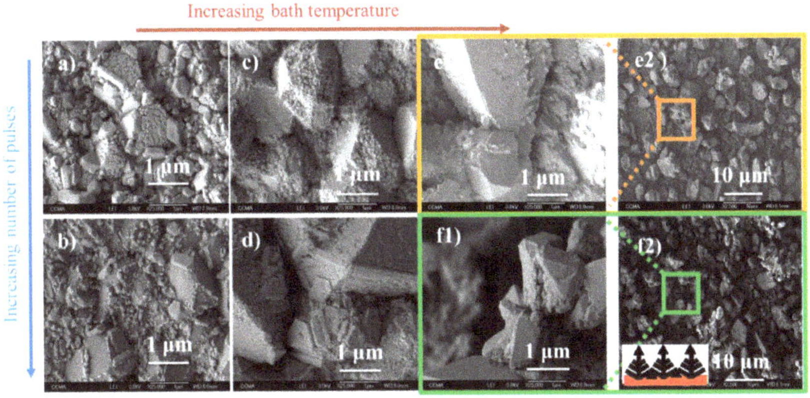

Figure 8. SEM images of the samples prepared from fluoroborate precursor using CV method. (**a**) 22 °C, 3 cycles, D_g: 200–700 nm on 1–1.2 μm grains; (**b**) 22 °C, 5 cycles, $D_g \approx$ 1–1.6 μm; (**c**) 45 °C, 3 cycles, $D_g \approx$ 1.5–2.5 μm; (**d**) 45 °C, 5 cycles, $D_g \approx$ 4–5 μm; (**e1**) 60 °C, 3 cycles, $D_g \approx$ 3–5.5 μm; (**e2**) a larger view of (**e1**); (**f1**) 60 °C, 5 cycles, with average tip sizes of 0.3–2 μm, in a larger view in (**f2**) with fractal trees length >10 μm, and a schematic pattern of the grown structure on the surface.

3.3. Wetting Durability

3.3.1. Under UV Exposure

To investigate the durability of wetting on the prepared copper thin layers, three samples with different surface structures, including micrometric crystals, dendrites, and hierarchical structures, were chosen (details in Figure 9).

Figure 9. SEM image of the samples chosen for the comparison between samples prepared by sulphate and fluoroborate precursors: (**a**) 'CuSO$_4$, 15 °C, 7 cycles', Cu(111) = 0.93%, Cu$_2$O (111) = 0.48%, $R_{a,O}$ = 806 ± 41 nm, crystal size < 1 μm, θ_S = 142°; (**b**) 'CuSO$_4$, 60 °C, 12 pulses', Cu(111) = 2.33%, Cu$_2$O (111) = 3.23%, $R_{a,O}$ = 5260 ± 387 nm, crystal size < 1 μm, θ_S = 160°, SA = 22°; (**c**) Cu(BF$_4$)$_2$, 60 °C, 12 pulses, Cu(111) = 0.92%, Cu$_2$O in (111) = 0.99%, $R_{a,O}$ = 5877 ± 747 nm; crystal size < 0.6 μm, θ_S = 160°, SA = 15°.

As discussed in our previous papers [23,27–29,38], hydrophobicity of the copper samples increases by aging in a sealed glass bottle. This change in contact angle could be a result of the contamination of sample with hydrocarbons in air, which is higher in the more-structured surfaces, i.e., dendrite and hierarchical surfaces, while the contact angle of as-prepared 'CuSO$_4$, 15 °C, 7 cycles', 'CuSO$_4$, 60 °C, 12 pulses', and 'Cu(BF$_4$)$_2$, 60 °C, 12 pulses' samples are 142°, 160°, and 160°, respectively (see Figure 9). Among these three samples, only 'CuSO$_4$, 60 °C, 12 pulses' shows good superhydrophobicity, and the water drop bounces on it (see Figure 10). Accordingly, during 200 ms after dropping a water drop, 1.36 mm in diameter, on the surface from 2.28 mm distance, the drop bounces six times before depositing on the substrate.

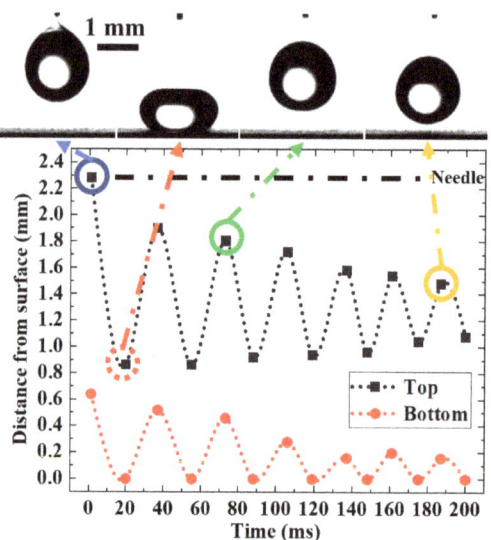

Figure 10. Six cycles of bounding, rebounding, bouncing of water drop on 'CuSO₄, 60 °C, 12 pulses' surface. Needle distance from the surface = 2.28 mm. Drop diameter = 1.36 mm.

According to the previous studies [47–49], superhydrophobicity in some metals is a result of the hydrocarbon adsorption on the surface from air and is not an intrinsic property. Hence, it is expected that UV irradiation cleans the surface from contaminations [50,51] and reduces its hydrophobicity. To study this effect, the wettability evolution under UV exposure is tracked up to 150 minutes and the results are shown in Figure 11. The highly hierarchical 'CuSO₄, 60 °C, 12 pulses' sample is still superhydrophobic even after 150 minutes: the contact angle is reduced from 160° to 145° during the initial 30 minutes of UV exposure and subsequently remains constant. As such, the surface is intrinsically superhydrophobic, and hydrocarbon adsorption does not play any role on the observed wetting behaviour of such surfaces. Differently, on 'CuSO₄, 15 °C, 7 cycles' and 'Cu(BF₄)₂, 60 °C, 12 pulses' the contact angle is reduced to ~125° in the first 70 minutes of UV exposure, and then remains constant. Thus, the samples are still hydrophobic, but not superhydrophobic anymore, suggesting that on these two surfaces hydrocarbon adsorption partially plays a role in conferring the initially observed superhydrophobicity. In order to study the wetting hysteresis in the exposed samples, advancing and receding contact angles of these samples were also measured.

According to Figure 12, the contact angle in 'CuSO₄, 15 °C, 7 cycles', 'CuSO₄, 60 °C, 12 pulses', and 'Cu(BF₄)₂, 60 °C, 12 pulses' is increased from 123°, 142°, and 128° just after the irradiation to 128°, 160°, and 142° after one week rest in the sealed glass bottle, respectively. Meanwhile, two less hydrophobic samples show a large hysteresis with a very low receding contact angle (i.e., $20° < \theta_{Rec} < 30°$), while 'CuSO₄, 60 °C, 12 pulses' is highly nonwettable with 6° wetting hysteresis. As a result of this study, the 'CuSO₄, 60 °C, 12 pulses' sample with a highly textured hierarchical surface is intrinsically superhydrophobic and is not highly affected by environmental contamination and UV irradiation. However, in the two less-structured surfaces, superhydrophobicity is less stable after cleaning with UV light, and superhydrophobicity in these two samples could be a result of the surface contamination.

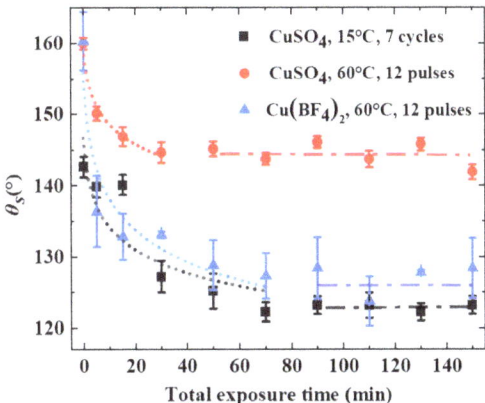

Figure 11. Changes in contact angle after UV exposure with 150 W power at 20 cm distance. Dotted lines show a power fitting to the data in reduction regime. Dash-dot lines represent the almost constant contact angles after reduction regime.

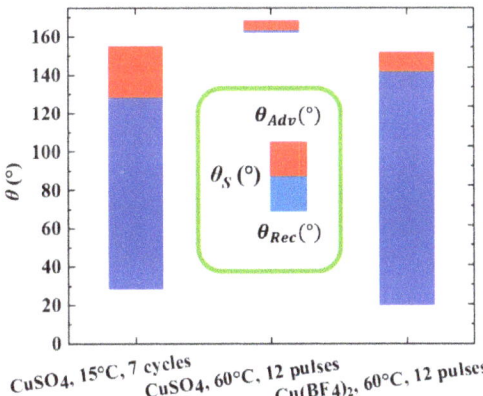

Figure 12. Static, advancing, and receding contact angles of the selected samples two days after UV exposure.

3.3.2. Abrasion Tests

In order to evaluate the stability of wetting in the selected highly rough fluoroborate and sulphate samples (i.e., 'Cu(BF$_4$)$_2$, 60 °C, 12 pulses' and 'CuSO$_4$, 60 °C, 12 pulses') under abrasion with sandpaper, the following test based on the ASTM D4060 standard [52] was performed.

As discussed in the supplementary of [38], the size of surface grooves after abrasion depends on the hardness of the surface structure, which influences the wetting stability of the sample in real harsh environments such as car or aircraft outside surfaces. According to Figure 13, after 30 cm hard abrasion, deep grooves with 6 and 18 μm width on average are created on the sulphate and fluoroborate samples, respectively. In a real application, the static contact angle and wetting hysteresis (i.e., $H = \theta_{Adv} - \theta_{Rec}$) are important.

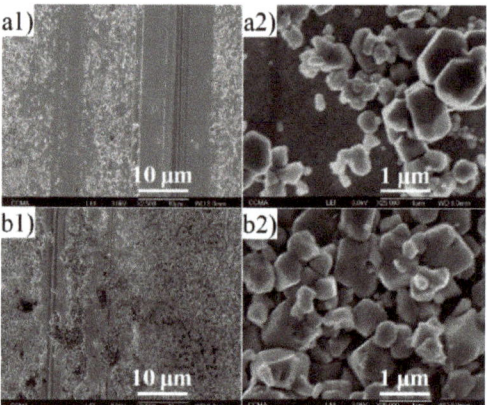

Figure 13. SEM images of (**a1,a2**) 'Cu(BF$_4$)$_2$, 60 °C, 12 pulses'; and (**b1,b2**) 'CuSO$_4$, 60 °C, 12 pulses', after 10-cycle abrasion with P1500 SiC sandpaper.

As shown in Figure 14, while the wetting state in the sulphate sample after a hard abrasion for 10 cycles is changed by 20° with maximum 6° hysteresis and is still highly hydrophobic, the fluoroborate sample faces a reduction in contact angles of ~40°. Moreover, while the sulphate samples remain constant in abrasion lengths larger than 12 cm, the wettability of the fluoroborate sample decreases even after 30 cm abrasion. Therefore, although the metal oxide surface will be destroyed after a hard abrasion with SiC sandpaper on the sulphate sample, the surface is still hydrophobic and shows a good wear resistance and is still reliable for practical applications under harsh environmental conditions.

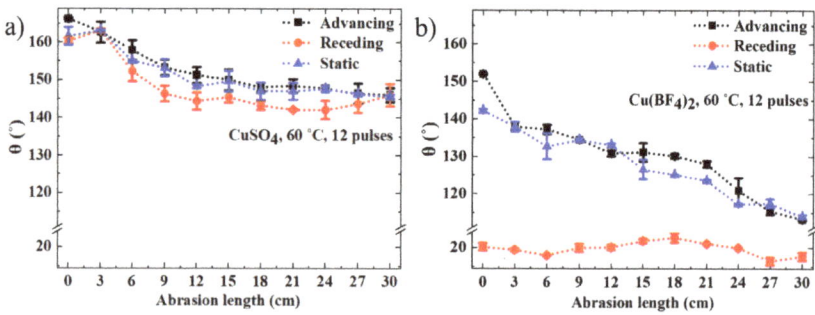

Figure 14. Changes in the advancing, receding, and static contact angles of (**a**) 'CuSO$_4$, 60 °C, 12 pulses' and (**b**) 'Cu(BF$_4$)$_2$, 60 °C, 12 pulses' samples under a hard abrasion for 10 cycles.

3.4. A Comparison between the Sulphate and Fluoroborate Samples

According to the present and our previously published works [29,38] on the fabrication of superhydrophobic copper surfaces using the electrodeposition method, the surface structures formed in the electrodeposition of copper using sulphate and fluoroborate precursors include (a) 3D crystals with flat facets in triangular, square and hexahedron shapes; (b) 100 nm balls grown on a coverage of micrometric octahedral crystals; (c) 10 to 20 μm fractal leaves with complex shapes and multiple branches and trunks, particularly with sharp octahedral crystals at the branch tips; and (d) a hierarchical micrometric combination of the previous structures. Accordingly, a comparison between the copper fluoroborate and sulphate prepared samples can be conducted as follows. Our experiments show that the maximum applicable bath temperature for aqueous solution is 65 and 90 °C [29] in fluoroborate and sulphate, respectively. Through the roughness curves, Figures 4 and 7, it could be concluded that the deposition rate of sulphate is higher than fluoroborate. A more careful study on both the lateral and vertical surface morphology is needed to approve this hypothesis. In the sense of chemical composition, a partial creation of the oxide grains after deposition under the air exposure due to higher porosity of the samples prepared in higher temperatures results a near decrease in the height of Cu(111) and increase in Cu_2O(111) by increasing the bath temperature in the samples prepared by the square-pulse method from sulphate and fluoroborate precursors, while the amount of copper oxide is still significantly lower than copper (Figure 3). Contrary to pulse samples, in cyclic voltammetry samples the XRD patterns (Figure 6) show the stability of copper in the system and the height of the Cu(111) peak is effectively increased by increasing the bath temperature, which could be a sign of the rapid increase in the layer thickness. Surface morphology and wetting studies show that increasing the bath temperature in the pulse voltage method increases the roughness and porosity rapidly by growing hierarchical and dendrite leaves on the surface of the copper samples, whereas the surface reaches the optimum structure and shows superhydrophobicity with $\theta_S > 150°$ and SA < 25° in an optical roughness (i.e., $R_{a,O}$) of around 5000 nm at 60 °C in both the sulphate and fluoroborate precursors (see Figure 4a,b). Nonetheless, the rapid increase in the roughness happens at 45 and 60 °C in sulphate fluoroborate samples, respectively. SEM images (Figure 7 of [38] and Figure 5) confirm the increase in the surface complexity in samples prepared from the both precursors, while the growth of a fully hierarchical surface structure with sharp tips and very low solid fraction in contact with the water drop in samples prepared at 60 °C provides enough qualifications for the creation of Cassie–Baxter conditions on the surface. The evolution rate of the surface structure in fluoroborate samples by increasing the bath temperature is lower than the sulphate samples. While the roughness and the contact angle in the fluoroborate samples prepared by cyclic voltammetry method has a similar trend, in samples prepared from sulphate precursor, the sample with higher roughness shows a lower contact angle (Figure 7) due to the powdery nature of the sample, and the water spreads over powder particles on the surface.

Durability tests of the samples on UV exposure and hard abrasion show the instability of the wetting state in the fluoroborate sample and its changes in the various environmental conditions, including UV light exposure and hard abrasion (Figures 11 and 14). Thus, the wetting state of the sulphate and fluoroborate samples is different, although these samples show a similar static contact angle and sliding angle in the as-prepared samples. This result emphasizes the importance of quasi-static wetting studies and systematic durability investigations in the superhydrophobic samples, which are going to be more noticed by scientists [53].

4. Conclusions

The present paper is an experimental study on the fabrication of robust superhydrophobic surfaces from copper-based precursors using electrodeposition. Specifically, both square-pulse voltage and cyclic voltammetry methods from copper fluoroborate precursor at different bath temperatures have been used. It has been observed that Cu and Cu_2O content on the deposited layer was affected by both the deposition temperature and the applied method. The increase in the bath temperature increased the roughness and water contact angle on the surface. Superhydrophobicity, with a contact angle of 160° and a sliding angle of 15°, was observed in the most structured surface fabricated by using pulse electrodeposition at 60 °C with 12 pulses: such sample has a hierarchical structure, including fractal leaves and submicron crystals. This sample was compared with a superhydrophobic sample fabricated from copper sulphate precursor, at the same deposition conditions, to investigate their durability under UV exposure and hard abrasion. It has been observed that on fluoroborate samples superhydrophobicity is not sustained, suggesting that hydrocarbon spontaneous adsorption from the atmosphere (which is removed by UV exposure) partially plays a role in conferring the initially observed superhydrophobicity. Differently, surfaces from sulphate precursor can sustain both UV exposure and hard abrasion, suggesting that they are intrinsically superhydrophobic, with no effect of hydrocarbon adsorption, and they can thus be more robust for industrial applications.

Author Contributions: Conceptualization, R.A., C.A. and T.D.; methodology, R.A., C.A. and T.D.; validation, R.A.; investigation, R.A.; resources, M.R.M., C.A. and T.D.; writing—original draft preparation, R.A.; writing—review and editing, R.A., M.R.M., C.A. and T.D.; visualization, R.A.; supervision, M.R.M., C.A., F.G. and T.D.; project administration, M.R.M., C.A., F.G. and T.D.; funding acquisition, R.A., M.R.M., C.A., F.G. and T.D. All authors have read and agreed to the published version of the manuscript.

Funding: Partial financial support by the Research Council of the University of Tehran is acknowledged. The authors would like to thank the Centre Commun de Microscopie Appliquée (CCMA), Nice, France, and University of Milano-Bicocca, especially Guilhem Godeau and Paolo Gentile for their support with SEM images. Financial support from the Italian Ministry of University and Research (MIUR) through grant "Dipartimenti di Eccellenza—2017 Materials For Energy" is gratefully acknowledged.

Institutional Review Board Statement: Not applicable.

Informed Consent Statement: Not applicable.

Data Availability Statement: Data is contained within the article.

Conflicts of Interest: The authors declare no conflict of interest.

References

1. Liu, J.; Huang, X.; Li, Y.; Li, Z.; Chi, Q.; Li, G. Formation of Hierarchical CuO Microcabbages as Stable Bionic Superhydrophobic Materials via a Room-Temperature Solution-Immersion Process. *Solid State Sci.* **2008**, *10*, 1568–1576. [CrossRef]
2. Zhang, P.; Lv, F.Y. A Review of the Recent Advances in Superhydrophobic Surfaces and the Emerging Energy-Related Applications. *Energy* **2015**, *82*, 1068–1087. [CrossRef]
3. Guittard, F.; Darmanin, T. *Bioinspired Superhydrophobic Surfaces: Advances and Applications with Metallic and Inorganic Materials*; Pan Stanford Publishing Pte Ltd, Ed.; Jenny Stanford Publishing: Stanford, CA, USA, 2017; ISBN 9781351859592.
4. Yifan Si, Z.G. Superhydrophobic Nanocoatings: From Materials to Fabrications and to Applications. *Nanoscale* **2015**, *7*, 5922–5946. [CrossRef]
5. Darmanin, T.; Guittard, F. Recent Advances in the Potential Applications of Bioinspired Superhydrophobic Materials. *J. Mater. Chem. A* **2014**, *2*, 16319–16359. [CrossRef]
6. Cassie, A.B.D.; Baxter, S. Wettability of Porous Surfaces. *Trans. Faraday Soc.* **1944**, *40*, 546–551. [CrossRef]
7. Khaskhoussi, A.; Risitano, G.; Calabrese, L.; D'andrea, D. Investigation of the Wettability Properties of Different Textured Lead/Lead-Free Bronze Coatings. *Lubricants* **2022**, *10*, 82. [CrossRef]
8. Volpe, A.; Covella, S.; Gaudiuso, C.; Ancona, A. Improving the Laser Texture Strategy to Get Superhydrophobic Aluminum Alloy Surfaces. *Coatings* **2021**, *11*, 369. [CrossRef]
9. Marmur, A. Hydro- Hygro- Oleo- Omni-Phobic? Terminology of Wettability Classification. *Soft Matter* **2012**, *8*, 6867–6870. [CrossRef]
10. Wenzel, R.N. Resistance of Solid Surfaces to Wetting by Water. *Ind. Eng. Chem.* **1936**, *28*, 988–994. [CrossRef]
11. Mumm, F.; Van Helvoort, A.T.J.; Sikorski, P. Easy Route to Superhydrophobic Copper-Based Wire-Guided Droplet Microfluidic Systems. *ACS Nano* **2009**, *3*, 2647–2652. [CrossRef]
12. Shirtcliffe, N.J.; McHale, G.; Newton, M.I.; Perry, C.C. Wetting and Wetting Transitions on Copper-Based Super-Hydrophobic Surfaces. *Langmuir* **2005**, *21*, 937–943. [CrossRef] [PubMed]
13. Wang, S.; Xue, Y.; Xue, Y.; Lv, C.; Jin, Y. Long-Term Durability of Robust Super-Hydrophobic Co–Ni-Based Coatings Produced by Electrochemical Deposition. *Coatings* **2022**, *12*, 222. [CrossRef]
14. Ramos Chagas, G.; Akbari, R.; Godeau, G.; Mohammadizadeh, M.; Guittard, F.; Darmanin, T. Electrodeposited Poly(Thieno[3,2-b]Thiophene) Films for the Templateless Formation of Porous Structures by Galvanostatic and Pulse Deposition. *Chempluschem* **2017**, *82*, 1351–1358. [CrossRef] [PubMed]
15. Darmanin, T.; De Givenchy, E.T.; Amigoni, S.; Guittard, F. Superhydrophobic Surfaces by Electrochemical Processes. *Adv. Mater.* **2013**, *25*, 1378–1394. [CrossRef]
16. Al-Bat'hi, S.A.M. Electrodeposition of Nanostructure Materials. In *Electroplating of Nanostructures*; Aliofkhazraei, M., Ed.; IntechOpen: London, UK, 2015; pp. 3–26.
17. Nasirpouri, F. *Electrodeposition of Nanostructured Materials*; Car, R., Ertl, G., Freund, H.J., Lüth, H., Rocca, M.A., Eds.; Springer: Cham, Switzerland, 2017; ISBN 9783319449197.
18. Gurrappa, I.; Binder, L. Electrodeposition of Nanostructured Coatings and Their Characterization—A Review. *Sci. Technol. Adv. Mater.* **2008**, *9*. [CrossRef]

19. Barnes, S.C.; Storey, G.G.; Pick, H.J. The Structure of Electrodeposited Copper-III. The Effect of Current Density and Temperature on Growth Habit. *Electrochim. Acta* **1960**, *2*, 195–204. [CrossRef]
20. Huang, M.C.; Wang, T.; Chang, W.S.; Lin, J.C.; Wu, C.C.; Chen, I.C.; Peng, K.C.; Lee, S.W. Temperature Dependence on P-Cu_2O Thin Film Electrochemically Deposited onto Copper Substrate. *Appl. Surf. Sci.* **2014**, *301*, 369–377. [CrossRef]
21. Mallik, A.; Ray, B.C. Implication of Low Temperature and Sonication on Electrocrystallization Mechanism of Cu Thin Films: A Kinetics and Structural Correlation. *Mater. Res.* **2013**, *16*, 539–545. [CrossRef]
22. Mallik, A.; Ray, B.C. Evolution of Principle and Practice of Electrodeposited Thin Film: A Review on Effect of Temperature and Sonication. *Int. J. Electrochem.* **2011**, *2011*, 568023. [CrossRef]
23. Akbari, R.; Mohammadizadeh, M.R.; Khajeh Aminian, M.; Abbasnejad, M. Hydrophobic Cu_2O Surfaces Prepared by Chemical Bath Deposition Method. *Appl. Phys. A Mater. Sci. Process.* **2019**, *125*, 190. [CrossRef]
24. Dini, J.W.; Snyder, D.D. Electrodeposition of Copper. In *Modern Electroplating*; Schlesinger, M.P., Ed.; John Wiley & Sons, Inc.: Hoboken, NJ, USA, 2010; pp. 33–78. ISBN 9780470167786.
25. Zhao, W.; Fu, W.; Yang, H.; Tian, C.; Li, M.; Li, Y.; Zhang, L.; Sui, Y.; Zhou, X.; Chen, H.; et al. Electrodeposition of Cu_2O Films and Their Photoelectrochemical Properties. *CrystEngComm* **2011**, *13*, 2871–2877. [CrossRef]
26. Ding, Y.; Li, Y.; Yang, L.; Li, Z.; Xin, W.; Liu, X.; Pan, L.; Zhao, J. The Fabrication of Controlled Coral-like Cu_2O Films and Their Hydrophobic Property. *Appl. Surf. Sci.* **2013**, *266*, 395–399. [CrossRef]
27. Akbari, R.; Ramos Chagas, G.; Godeau, G.; Mohammadizadeh, M.; Guittard, F.; Darmanin, T. Intrinsically Water-Repellent Copper Oxide Surfaces; An Electro-Crystallization Approach. *Appl. Surf. Sci.* **2018**, *443*, 191–197. [CrossRef]
28. Akbari, R.; Godeau, G.; Mohammadizadeh, M.; Guittard, F.; Darmanin, T. Wetting Transition from Hydrophilic to Superhydrophobic over Dendrite Copper Leaves Grown on Steel Meshes. *J. Bionic Eng.* **2019**, *16*, 719–729. [CrossRef]
29. Akbari, R.; Godeau, G.; Mohammadizadeh, M.; Guittard, F.; Darmanin, T. The Influence of Bath Temperature on the One-Step Electrodeposition of Non-Wetting Copper Oxide Coatings. *Appl. Surf. Sci.* **2020**, *503*, 144094. [CrossRef]
30. Sau, T.K.; Rogach, A.L. *Complex-Shaped Metal Nanoparticles*; Sau, T.K., Rogach, A.L., Eds.; Wiley-VCH: Weinheim, Germany, 2012; ISBN 9783527325894.
31. Ko, W.Y.; Chen, W.H.; Der Tzeng, S.; Gwo, S.; Lin, K.J. Synthesis of Pyramidal Copper Nanoparticles on Gold Substrate. *Chem. Mater.* **2006**, *18*, 6097–6099. [CrossRef]
32. Ko, W.Y.; Chen, W.H.; Cheng, C.Y.; Lin, K.J. Highly Electrocatalytic Reduction of Nitrite Ions on a Copper Nanoparticles Thin Film. *Sens. Actuators B Chem.* **2009**, *137*, 437–441. [CrossRef]
33. Ko, W.Y.; Chen, W.H.; Cheng, C.Y.; Lin, K.J. Architectural Growth of Cu Nanoparticles through Electrodeposition. *Nanoscale Res. Lett.* **2009**, *4*, 1481–1485. [CrossRef]
34. Wang, L. *Preparation and Characterization of Properties of Electrodeposited Copper Oxide Films*; The University of Texas at Arlington: Arlington, TX, USA, 2006.
35. Giri, S.D.; Sarkar, A. Electrochemical Study of Bulk and Monolayer Copper in Alkaline Solution. *J. Electrochem. Soc.* **2016**, *163*, H252–H259. [CrossRef]
36. Kim, J.; Kim, Y.; Jung, J.; Chae, W.S. Photoassisted Electrodeposition of a Copper(I) Oxide Film. *Mater. Trans.* **2015**, *56*, 377–380. [CrossRef]
37. Free, M.; Rodchanarowan, A.; Phadke, N.; Bhide, R. Evaluation of the Effects of Additives, Pulsing, and Temperature on Morphologies of Copper Electrodeposited from Chloride Media. *ECS Trans.* **2006**, *2*, 335–343. [CrossRef]
38. Akbari, R.; Godeau, G.; Mohammadizadeh, M.R.; Guittard, F.; Darmanin, T. Fabrication of Superhydrophobic Hierarchical Surfaces by Square Pulse Electrodeposition: Copper-Based Layers on Gold/Silicon (100) Substrates. *Chempluschem* **2019**, *84*, 368–373. [CrossRef] [PubMed]
39. Chen, L.-C.; Nguyen, D.T.; Chang, Y.-W. Precise Optical Surface Profilometry Using Innovative Chromatic Differential Confocal Microscopy. *Opt. Lett.* **2016**, *41*, 5660. [CrossRef] [PubMed]
40. Vorburger, T.V.; Rhee, H.G.; Renegar, T.B.; Song, J.F.; Zheng, A. Comparison of Optical and Stylus Methods for Measurement of Surface Texture. *Int. J. Adv. Manuf. Technol.* **2007**, *33*, 110–118. [CrossRef]
41. Chand, M.; Mehta, A.; Sharma, R.; Ojha, V.N.; Chaudhary, K.P. Roughness Measurement Using Optical Profiler with Self-Reference Laser and Stylus Instrument—A Comparative Study. *Indian J. Pure Appl. Phys.* **2011**, *49*, 335–339.
42. Akbari, R.; Antonini, C. Contact Angle Measurements: From Existing Methods to an Open-Source Tool. *Adv. Colloid Interface Sci.* **2021**, *294*, 102470. [CrossRef]
43. Siegfried, M.J.; Choi, K.S. Electrochemical Crystallization of Cuprous Oxide with Systematic Shape Evolution. *Adv. Mater.* **2004**, *16*, 1743–1746. [CrossRef]
44. Siegfried, M.J.; Choi, K.-S. Directing the Architecture of Cuprous Oxide Crystals during Electrochemical Growth. *Angew. Chem.* **2005**, *117*, 3282–3287. [CrossRef]
45. Siegfried, M.J.; Choi, K.S. Elucidating the Effect of Additives on the Growth and Stability of Cu_2O Surfaces via Shape Transformation of Pre-Grown Crystals. *J. Am. Chem. Soc.* **2006**, *128*, 10356–10357. [CrossRef]
46. Siegfried, M.J.; Choi, K.S. Elucidation of an Overpotential-Limited Branching Phenomenon Observed during the Electrocrystallization of Cuprous Oxide. *Angew. Chem.-Int. Ed.* **2008**, *47*, 368–372. [CrossRef]
47. Tam, J.; Palumbo, G.; Erb, U.; Azimi, G. Robust Hydrophobic Rare Earth Oxide Composite Electrodeposits. *Adv. Mater. Interfaces* **2017**, *4*, 1700850. [CrossRef]

48. Hassebrook, A.C. *Applications of Femtosecond Laser Processed Metallic Surfaces: Leidenfrost Point and Thermal Stability of Rare Earth Oxide Coatings*; University of Nebraska-Lincoln: Lincoln, NE, USA, 2017.
49. Preston, D.J.; Miljkovic, N.; Sack, J.; Enright, R.; Queeney, J.; Wang, E.N. Effect of Hydrocarbon Adsorption on the Wettability of Rare Earth Oxide Ceramics Effect of Hydrocarbon Adsorption on the Wettability of Rare Earth Oxide Ceramics. *Appl. Phys. Lett.* **2014**, *105*, 011601. [CrossRef]
50. Boyce, J.M. Modern Technologies for Improving Cleaning and Disinfection of Environmental Surfaces in Hospitals. *Antimicrob. Resist. Infect. Control* **2016**, *5*, 10. [CrossRef] [PubMed]
51. González, C.M. Cleaning with UV Light. *Mech. Eng.* **2021**, *143*, 32–33. [CrossRef]
52. *ASTM D 4060-10*; Standard Test Method for Abrasion Resistance of Organic Coatings by the Taber. ASTM International: West Conshohocken, PA, USA, 2010. [CrossRef]
53. Zhang, Y.; Liu, J.; Ouyang, L.; Li, J.; Xie, G.; Yan, Y.; Weng, C. One-Step Preparation of Robust Superhydrophobic Foam for Oil/Water Separation by Pulse Electrodeposition. *Langmuir* **2021**, *37*, 7043–7054. [CrossRef]

Article

Candle Soot-Based Electrosprayed Superhydrophobic Coatings for Self-Cleaning, Anti-Corrosion and Oil/Water Separation

Yuting Zhang [1], Tingping Lei [1,2,*], Shuangmin Li [1], Xiaomei Cai [3,*], Zhiyuan Hu [1], Weibin Wu [3] and Tianliang Lin [1]

1. College of Mechanical Engineering and Automation, Huaqiao University, Xiamen 361021, China; ytzhang@stu.hqu.edu.cn (Y.Z.); 1711113012@stu.hqu.edu.cn (S.L.); hzy1999@mail.ustc.edu.cn (Z.H.); ltl@hqu.edu.cn (T.L.)
2. Fujian Provincial Key Laboratory of Special Energy Manufacturing, Huaqiao University, Xiamen 361021, China
3. School of Science, Jimei University, Xiamen 361021, China; 201921145023@jmu.edu.cn
* Correspondence: tplei@hqu.edu.cn (T.L.); cxplum@163.com (X.C.)

Abstract: The interest in candle soot (CS)-based superhydrophobic coatings has grown rapidly in recent years. Here, a simple and low-cost process has been developed for the fabrication of CS-based superhydrophobic coatings through electrospraying of the composite cocktail solution of CS and polyvinylidene fluoride (PVDF). Results show that the superhydrophobicity of the coating closely relates to the loading amount of CS which results in coatings with different roughnesses. Specifically, increasing the CS amount (not more than 0.4 g) normally enhances the superhydrophobicity of the coating due to higher roughness being presented in the produced microspheres. Further experiments demonstrate that the superhydrophobicity induced in the electrosprayed coating results from the synergistic effect of the cocktail solution and electrospray process, indicating the importance of the coating technique and the solution used. Versatile applications of CS-based superhydrophobic coatings including self-cleaning, anti-corrosion and oil/water separation are demonstrated. The present work provides a convenient method for the fabrication of CS-based superhydrophobic coatings, which is believed to gain great interest in the future.

Keywords: superhydrophobic coatings; electrospraying; candle soot; self-cleaning; anti-corrosion; oil/water separation

1. Introduction

Candle soot (CS), traditionally deemed as a source of unwanted air pollution, has received increasing attention in recent years [1–3]. Fresh CS collected from the inner flame is superhydrophobic (water contact angle, WCA $\geq 150°$) [4], but pristine CS is fragile and oxidation during aging causes the soot to become hydrophilic [5]. To obviate these problems, researchers have proposed numerous approaches that can be classified into the following three major categories: (1) substrate pretreatment before CS deposition, either through coating polydimethylsiloxane (PDMS) mixtures [6–8] or paraffin wax [9] on the raw substrate, or making the substrate much rougher (e.g., via electrodeposition [10] or other methods [11,12]); (2) reinforcement after CS deposition, mainly via covering the CS layer with PDMS mixtures [13,14] or some specific polymer solutions [15]; and (3) mixing CS with polymer for solution deposition of superhydrophobic coatings [16–18].

Among them, the method of mixing CS with polymer is much simpler and more flexible in the selection of both polymer materials and deposition methods. The materials, such as PDMS, polyurethane (PU), and polyvinylidene fluoride (PVDF), and the deposition methods, such as spray coating, spin coating, and gelation technique, can be used for making superhydrophobic polymer/CS composite coatings. Literature suggests that spray coating has been widely adopted by many researchers. Sutar et al. reported the use of the spray technique to deposit PDMS/CS composite for self-cleaning superhydrophobic coating [19]. Li et al. employed the spray technique to fabricate superhydrophobic PU/CS coatings for oil–water separation [16]. The spray coating was also used for the scalable fabrication of superhydrophobic PVDF/SiO$_2$ membranes for gravitational water-in-oil emulsion separation [20]. Compared with the conventional spray coating, electrospraying (electrohydrodynamic spraying, a process utilizing the electric field alone rather than additional mechanical energy to generate fine droplets with charge) allows for better control and higher deposition efficiency of the atomized charged droplets to self-disperse in smaller sizes [21]. The technique of electrospraying has been intensively studied in the synthesis of micro/nano materials [22,23] and mass spectrometry [24,25]. To the best of our knowledge, electrospraying of CS-based superhydrophobic coatings are seldom reported, although recent works utilized this technique to deposit functionalized CS particles and carbonaceous nanoparticle layers [26,27].

In this work, CS-based superhydrophobic coatings were demonstrated through electrospraying of the composite cocktail solution of CS mixed with PVDF (a commercially available fluoropolymer with low surface energy, 25 dynes/cm). The synergetic effect and the main factors influencing the hydrophobicity of the coatings were investigated, and the typical applications in self-cleaning, anti-corrosion, and oil/water separation were presented.

2. Experimental

Polyvinylidene fluoride (M_W ~ 625,000) in powder form was purchased from Shanghai Sensure Chemical Co., Ltd. (Xi'an, China). Analytically pure N, N-dimethylformamide (DMF), and acetone were obtained from Sinopharm Chemical Reagent Co., Ltd. (Shanghai, China) and used as received. Candles and substrates including glass slides, stainless steel and copper meshes, cotton fabrics, printing paper, iron sheets, wood panels, and cobblestones were purchased from local supermarkets. The superhydrophobic CS was collected from the middle zone of candle flame as described previously [18,28]. Specifically, by placing a stainless steel plate above the outer flame of the burning candle for 1 min, a thick layer of CS particles was obtained on the plate, which could be further scraped and transferred for use.

A standard electrospray apparatus was utilized to deposit PVDF-CS composite coatings, where the solution flow rate was controlled by a precision syringe pump (NE-300, New Era Pump Systems Inc., Farmingdale, NY, USA). The cocktail solution was prepared by continuously stirring the mixture of 0.3 g CS (if not stated otherwise) with a PVDF solution that was formed by dissolving 0.2 g PVDF powders in 6 mL DMF and 4 mL acetone. Before the electrospraying process, the cocktail solution was degassed to remove air bubbles. During the electrospraying experiments, the applied voltage was set at 7.0 kV, the solution feed rate was 500 µL/h, and the needle tip-to-collector distance was 10 cm. All experiments were performed under an air atmosphere with a relative humidity between 55 and 60%.

The surface morphology of the as-prepared samples was observed by a scanning electron microscope (SEM, SU70 and SU5000, Hitachi, Tokyo, Japan). The hydrophobicity characterization was conducted on a contact angle analyzer (JC2000D3, Shanghai Zhongchen Digital Technology Equipment Co., Ltd., Shanghai, China) by placing water droplets of 9 µL on the coatings. The contact angle data were figured out based on the ellipse fitting method, and the final result was averaged from five measurements per specimen.

The self-cleaning property was evaluated by sprinkling the chalk powder on a tilted coated substrate and slowly dropping water droplets on it. The oil/water separation was performed by pouring oil/water mixtures into the coated copper mesh that was tailed into a "Taylor cone" container, where hexadecane and chloroform were used as light and heavy oils, respectively. The anti-corrosion was tested by soaking stainless steel mesh (both uncoated and coated) in concentrated HCl solution.

3. Results and Discussion

Figure 1 illustrates the hydrophobicity and morphology of the electrosprayed coatings on paper and Al foil substrates with different loading amounts of superhydrophobic CS. As shown in Figure 1A, with the loading of CS into PVDF solution, all PVDF/CS composite coatings either on paper or on the Al foil substrate show visible improvement in hydrophobicity (as compared with the pure PVDF coatings), although there are some differences. When 0.05 g CS is loaded, the coated paper already becomes superhydrophobic (average WCA ~ 153°) and the coated Al foil sample is also close to superhydrophobic (average WCA ~ 145°). Further increasing the CS loading allows the composite coatings to further improve in superhydrophobicity. The highest WCA obtained in our repeated experiments was about 168° for the coated paper when a higher amount of CS (e.g., 0.4 g) was loaded. The representative WCA photos of the corresponding curves in Figure 1A are demonstrated in Figure 1B, where the arrow indicates the increasing direction of CS content.

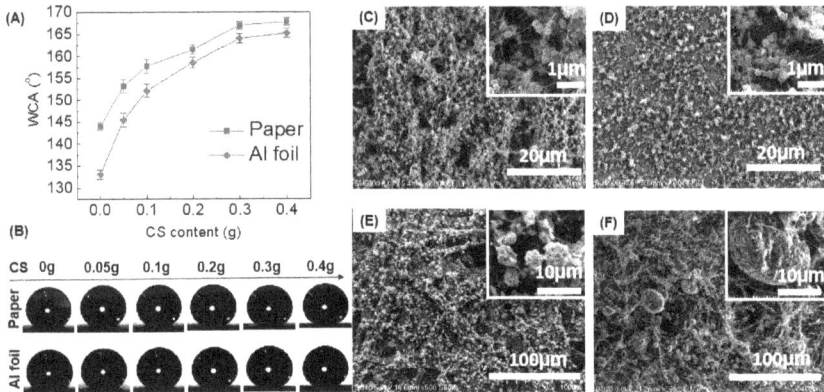

Figure 1. Effect of CS content on the hydrophobicity and surface morphology of electrosprayed paper and Al foil substrates: (**A**) water contact angle (WCA) varied with CS content; (**B**) typical WCA photos of the curves from (**A**) (the arrow indicates the increasing direction of CS content); (**C–F**) SEM images of electrosprayed samples with CS contents of 0 g and 0.3 g on paper (**C,E**) and Al foil (**D,F**), respectively (insets are the enlarged images of each samples).

Surface morphologies of the pure PVDF coatings and the typical PVDF/CS coatings on both substrates are shown in Figure 1C–F. It is observed that electrosprayed coatings from pure PVDF solution are chiefly composed of smooth nanospheres (200~300 nm), although with some ultrafine fibers (Figure 1C,D). In contrast, the coatings electrosprayed with the PVDF/CS cocktail solution present rough microspheres in large numbers (Figure 1E,F). That is why the electrosprayed coatings using the low surface energy materials of PVDF and CS show better hydrophobicity than the coatings using the pure PVDF. On the other hand, from the results shown in Figure 1A, it is reasonable to infer that increasing the CS loading enhances the roughness of the coatings, and therefore a better hydrophobicity is normally obtained. However, it should be noted that when the loading amount of CS reaches 0.4 g, continuous loading has little contribution to the enhancement of hydrophobicity.

Besides the hydrophobicity, the "stickiness" of the coatings is also significant from the application point of view. Figure 2A_1,A_2 shows that pure PVDF coatings (either on the Al foil or paper substrate) are very sticky, as evidenced by the observation of semi-spherical water droplets on the coatings at a tilt of 180°. This highly sticky hydrophobic coating was reported to be the result of the pseudo-hydrogen bonding effect of the polarized C-H bonds in each repeating unit of PVDF polymer chains [29]. Unlike pure PVDF coatings, PVDF/CS composite coatings show different stickiness behaviors for the substrate. The composite coatings on paper allow the water droplet to roll off at a tilt of 45° (Figure 2B$_2$), whereas the coatings on the Al foil still show water-stickiness (Figure 2B$_1$) although not so strong as the pure PVDF coatings in Figure 2A$_1$,A$_2$. Thus, the loading of CS into PVDF solution during the electrospraying process not only enhances the coating hydrophobicity, but also reduces the "stickiness" of the coatings, which is useful for potential applications in the field of self-cleaning and anti-corrosion.

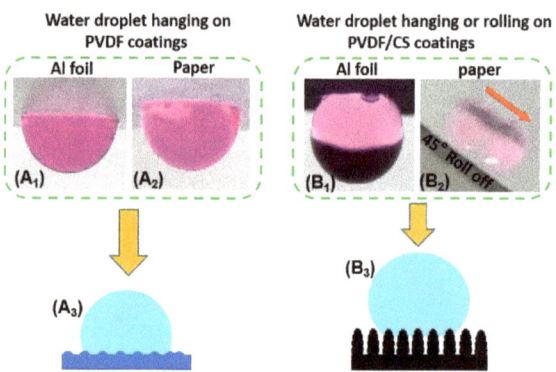

Figure 2. Wetting behavior and regime of the electrosprayed coatings: (**A$_1$,A$_2$**) and (**B$_1$**) water droplets hanging on the coatings at a tilt of 180°; (**B$_2$**) water droplet rolling off the coatings at a tilt of 45°; (**A$_3$,B$_3$**) schematics of the Wenzel model and the combined Cassie–Baxter/Wenzel model for the corresponding coatings, respectively.

According to the surface morphology of the coatings (Figure 1C–F) and their wetting behavior (hydrophobicity (Figure 1A) and "stickiness" (Figure 2A$_1$,A$_2$,B$_1$,B$_2$), it is reasonable to state that hydrophobic electrosprayed PVDF coatings follow the total wetting Wenzel state (Figure 2A$_3$), and superhydrophobic electrosprayed PVDF/CS coatings follow the combined Cassie–Baxter/Wenzel state (Figure 2B$_3$). It should be emphasized that the superhydrophobicity presented in the electrosprayed coatings is attributed to the synergistic effect of using the PVDF/CS cocktail solution and the process employed. As shown in Figure 1, the electrosprayed coatings from pure PVDF solution (without CS loading) are hydrophobic rather than superhydrophobic, since the assembly of smooth nanospheres can not make the coating rough enough to achieve superhydrophobicity [30]. In contrast, the electrosprayed coatings with the PVDF/CS cocktail solution easily become superhydrophobic due to the formation of numerous rough microspheres that probably result from incompatible CS nanoparticles and PVDF macromolecules bonded together at their contact points or interfaces (since CS and PVDF are of completely different solubilities in DMF). However, it is noted that the PVDF/CS coatings fabricated by spin coating or solution casting show poorer hydrophobicity than the electrosprayed PVDF coatings. In this regard, electrospraying is more powerful in the fabrication of (super)hydrophobic coatings.

In the electrospraying process, the solution is highly charged and undergoes Coulomb fission and breaks into tiny self-dispersing droplets [21]. As the solvent evaporates, deposits (coatings) of different morphologies and roughnesses are produced on the target substrate depending on the solution parameters (solution composition, solvent evaporation rate, etc.) and the processing parameters (applied voltage, solution feed rate, and the needle tip-to-collector distance, etc.). Thus, the coatings electrosprayed using (cocktail) solution with low surface energy are hydrophobic or superhydrophobic, as shown in Figure 1. This simple process also allows for superhydrophobic deposits on any solid substrates of various geometrical configurations [31,32]. The representative substrates (including glass slide, cotton fabric, printing paper, iron sheet, wood panel, and cobblestone) with PVDF/CS coatings show excellent water repellency as compared with the untreated substrates (Figure 3A). Moreover, a double-faced superhydrophobic printing paper was fabricated via electrospraying with PVDF/CS composites on both sides of the paper. As shown in Figure 3B, the coated paper is perfectly clean as it is taken out from the red water, whereas the uncoated paper is completely wetted by the water and leaves the stains on the cotton.

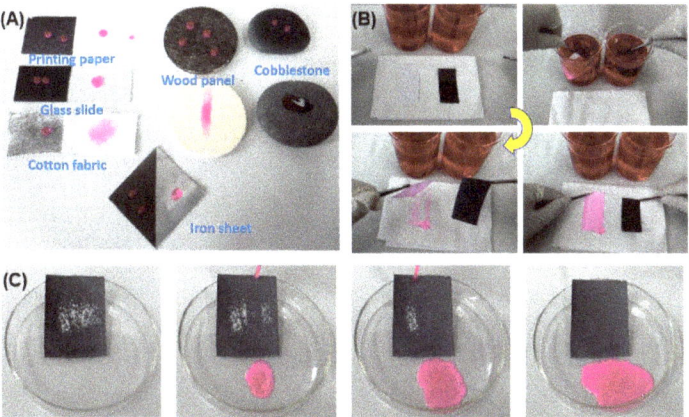

Figure 3. (**A**) Wetting behaviors of various substrates with/without electrosprayed PVDF/CS coatings by placing some 12 µL water droplets (dyed red) on them; (**B**) wetting behaviors of double-faced superhydrophobic printing paper (coated with the same coatings as (**A**)) and the uncoated paper, when they are first immersed into the red water and then placed on the cotton; (**C**) snapshots showing the self-cleaning property of the coated printing paper by slowly dropping water droplets on the surface which has been sprinkled with the chalk powder.

Nowadays, self-cleaning plays a significant role in various applications (such as buildings, solar panels and wind shields), in that keeping the surface self-clean means that no additional cost is required for the maintenance, and the service life is also increased. Figure 3C demonstrates the self-cleaning performance of electrosprayed PVDF/CS-coated printing paper by randomly sprinkling chalk powder (used as the contaminant) on the coated substrate that was tilted 15°. A red water droplet (12 µL) was slowly dropped on the substrate from a height of 1.5 cm above the surface. The powder particles are taken away with water droplets as they slide down the surface (Figure 3C), suggesting the self-cleaning nature.

The wetting property of the electrosprayed PVDF/CS coating was further investigated by subjecting it to droplets of various liquids. As shown in Figure 4, all aqueous droplets (H_2SO_4, KOH, and hot water) exhibit perfect spherical shapes, except for the milk droplet which slightly lost its sphericity (possibly due to the different composition in milk that contains protein, fat, lactose, and minerals besides water [33]), whereas the oil droplets spread over the coated paper. In fact, n-hexadecane (3.36 cP) spreads immediately as it touches the surface, indicating that the coating is also superoleophilic. On the other hand, the long-term superhydrophobicity to concentrated H_2SO_4 (pH = 1) and KOH (pH = 14) solutions suggests the feasibility of using the coating under harsh conditions. The above results show the potential for practical use of the coating in the separation of low viscosity oil/water mixtures.

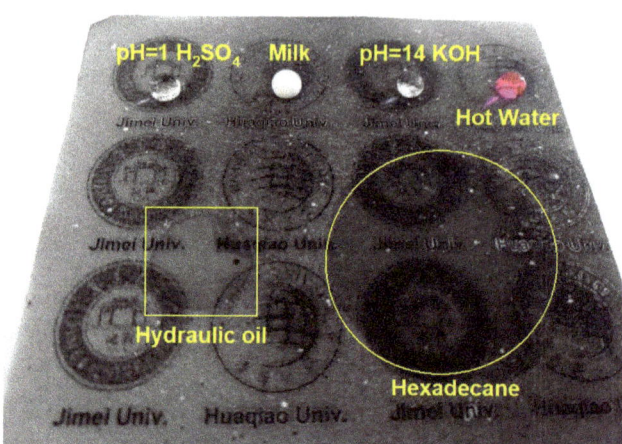

Figure 4. Super big droplets (c.a. 0.15 mL) of various liquids on the printing paper that is coated with electrosprayed PVDF/CS composites. Details are indicated in the figure, where the temperature for the hot water is about 85 °C.

As was previously pointed out by most publications [28,34,35], the use of superhydrophobic and superoleophilic (or superhydrophilic and underwater superoleophobic) mesh in oil/water separation normally encountered the problem of a water (or oil) barrier that blocks oil (or water) from passing through the membrane. To avoid this problem, we propose a "Taylor cone"-like 3D mesh container that allows both oil and water to touch the container wall, such that the separation of both light ($\rho_{oil} < \rho_{water}$) and heavy ($\rho_{oil} > \rho_{water}$) oil/water mixtures can be achieved. As illustrated in Figure 5A, when light oil/water mixtures are loaded into the superhydrophobic and superoleophilic mesh container, the oil (green) can pass the mesh even though water (red) settles down to form the barrier. For the separation of heavy oil/water mixtures, the oil (light blue) can directly pass through the bottom, whereas water (red) is blocked in the container. To validate the proposed model, a superhydrophobic and superoleophilic copper mesh (120) was electrosprayed with PVDF/CS composites, the SEM image of which is shown in Figure 5B. The as-prepared mesh was then sandwiched between two identical structural stainless steel meshes (200) to form the "Taylor cone" container. Light oil/water separation was performed by pouring a mixture of 10 mL hexadecane and 10 mL water into the container, and the separation was successfully finished after a few seconds (Figure 5C). A similar process was successfully conducted for heavy oil/water separation, where chloroform was used as the heavy oil (Figure 5D). By properly pumping out the water accumulated in the container, continuous oil/water separation can be realized for oils of any density [28].

The anti-corrosion experiment was conducted by immersing the uncoated stainless steel mesh and the double-faced superhydrophobic mesh (coated with PVDF/CS coatings) into the concentrated HCl solution. As shown in Figure 6, the uncoated mesh reacts with the solution more rapidly than the coated mesh, as revealed by darkening of the solution color. After 30 min, both of the meshes were carefully picked up for comparison. The double-faced mesh was nearly intact, whereas the uncoated mesh was badly destroyed, indicating the potential of using the coatings to protect the metal from corrosion. The agglomeration and chemical inertness of CS particles provide PVDF/CS coatings with a rough surface topography and chemical inhomogeneity, allowing the coatings to be in the Cassie–Baxter state to slow down solution erosion [11,36].

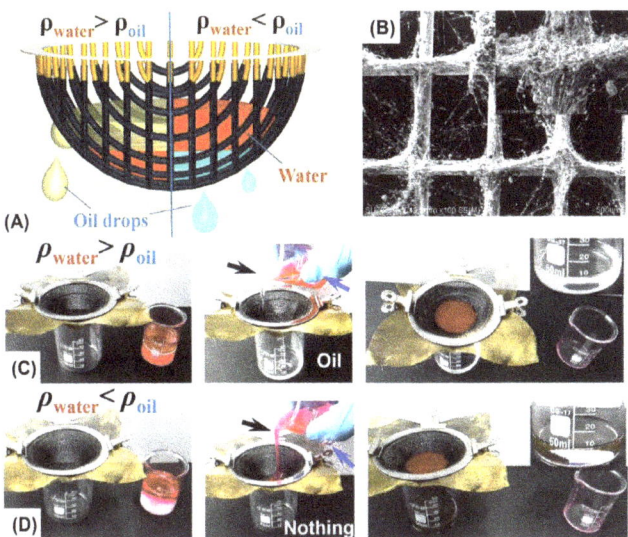

Figure 5. (**A**) Schematics of a superhydrophobic and superoleophilic "Taylor cone"-like 3D mesh container for both light and heavy oil/water separation; (**B**) SEM image of the coated copper mesh with electrosprayed PVDF/CS composites (enlarged in inset); snapshots of the use of the mesh container to separate light oil/water mixtures (**C**) and heavy oil/water mixtures (**D**), where hexadecane and chloroform are used as light and heavy oils, respectively. The water was dyed red.

Figure 6. Comparison of anti-corrosive behavior for the uncoated stainless steel mesh and the double-faced superhydrophobic mesh coated with PVDF/CS in concentrated HCl solution.

4. Conclusions

In summary, a simple and low-cost process for the fabrication of CS-based superhydrophobic coating was developed. Electrospraying with CS and PVDF composite cocktail solution can induce numerous rough microspheres in the coating and therefore realize superhydrophobicity. The comparison of the solution and coating methods reveals that the formation of superhydrophobicity is attributed to the synergistic effect of the cocktail solution and electrospraying. Both the CS loading and the substrate influence the superhydrophobicity of the CS-based electrosprayed coatings. The as-prepared CS-based coatings show potential applications in the fields of self-cleaning, oil/water separation, and anti-corrosion.

Author Contributions: Conceptualization, T.L. (Tingping Lei) and X.C.; methodology, Y.Z.; software, S.L.; validation, Y.Z., S.L. and Z.H.; formal analysis, W.W.; investigation, Y.Z.; resources, T.L. (Tingping Lei); data curation, W.W.; writing—original draft preparation, Y.Z.; writing—review and editing, T.L. (Tingping Lei); visualization, Y.Z. and Z.H.; supervision, T.L. (Tingping Lei); project administration, X.C., T.L. (Tingping Lei) and T.L. (Tianliang Lin); funding acquisition, X.C., T.L. (Tingping Lei) and T.L. (Tianliang Lin). All authors have read and agreed to the published version of the manuscript.

Funding: This work was supported by the Natural Science Foundation of Fujian Province (No. 2021J01298 and 2020J01709), the Youth Innovation Fund of Xiamen City (No. 3502Z20206010), and the Collaborative Innovation Platform of Fuzhou-Xiamen-Quanzhou Independent Innovation Demonstration Area (No. 3502ZCQXT202002).

Institutional Review Board Statement: Not applicable. The study did not require ethical approval.

Informed Consent Statement: Not applicable. The study did not involve humans.

Data Availability Statement: The data presented in this study are available in the manuscript.

Conflicts of Interest: The authors declare no conflict of interest.

References

1. Mulay, M.R.; Chauhan, A.; Patel, S. Viswanath Balakrishnan, Aditi Halder, Rahul Vaish, Candle soot: Journey from a pollutant to a functional material. *Carbon* **2019**, *144*, 684–712. [CrossRef]
2. Deng, X.; Mammen, L.; Butt, H.; Vollmer, D. Candle Soot as a Template for a Transparent Robust Superamphiphobic Coating. *Science* **2012**, *335*, 67–70. [CrossRef] [PubMed]
3. Liu, H.; Ye, T.; Mao, C. Fluorescent Carbon Nanoparticles Derived from Candle Soot. *Angew. Int. Ed.* **2007**, *46*, 6473–6475. [CrossRef] [PubMed]
4. Liang, C.; Liao, J.; Li, A.; Chen, C.; Lin, H.; Wang, X.; Xu, Y. Relationship between wettabilities and chemical compositions of candle soots. *Fuel* **2014**, *128*, 422–427. [CrossRef]
5. Zuberi, B.; Johnson, K.S.; Aleks, G.K.; Molina, L.T.; Laskin, A. Hydrophilic properties of aged soot. *Geophys. Res. Lett.* **2005**, *32*, L01807. [CrossRef]
6. Zhang, B.; Duan, J.; Huang, Y.; Hou, B. Double layered superhydrophobic PDMS-Candle soot coating with durable corrosion resistance and thermal-mechanical robustness. *J. Mater. Sci. Technol.* **2021**, *71*, 1–11. [CrossRef]
7. Liu, X.; Zhang, X.; Chen, Q.; Pan, Y.; Liu, C.; Shen, C. A simple superhydrophobic/superhydrophilic Janus-paper with enhanced biocompatibility by PDMS and candle soot coating for actuator. *Chem. Eng. J.* **2021**, *406*, 126532. [CrossRef]
8. Iqbal, R.; Majhy, B.; Sen, A.K. Facile Fabrication and Characterization of a PDMS-Derived Candle Soot Coated Stable Biocompatible Superhydrophobic and Superhemophobic Surface. *ACS Appl. Mater. Interfaces* **2017**, *9*, 31170–31180. [CrossRef]
9. Seo, K.; Kim, M.; Kim, D.H. Candle-based process for creating a stable superhydrophobic surface. *Carbon* **2014**, *68*, 583–596. [CrossRef]
10. Cao, H.; Fu, J.; Liu, Y.; Chen, S. Facile design of superhydrophobic and superoleophilic copper mesh assisted by candle soot for oil water separation. *Colloids Surf. A Physicochem. Eng. Asp.* **2018**, *537*, 294–302. [CrossRef]
11. Zhang, J.; Rosenkranz, A.; Zhang, J.; Guo, J.; Li, X.; Chen, X.; Xiao, J.; Xu, J. Modified wettability of micro-structured steel surfaces fabricated by elliptical vibration diamond cutting. *Int. J. Precis. Eng. Manuf.-Green Technol.* **2021**, 1–11. [CrossRef]
12. Barraza, B.; Olate-Moya, F.; Montecinos, G.; Ortega, J.H.; Rosenkranz, A.; Tamburrino, A.; Palza, H. Superhydrophobic SLA 3D printed materials modified with nanoparticles biomimicking the hierarchical structure of a rice leaf. *Sci. Technol. Adv. Mater.* **2022**, *23*, 300–321. [CrossRef] [PubMed]
13. Wu, S.; Du, Y.; Alsaid, Y.; Wu, D.; Hua, M.; Yan, Y.; Yao, B.; Ma, Y.; Zhu, X.; He, X. Superhydrophobic photothermal icephobic surfaces based on candle soot. *Proc. Natl. Acad. Sci. USA* **2020**, *117*, 11240–11246. [CrossRef]
14. Zhang, X.; Pan, Y.; Gao, Q.; Zhao, J.; Wang, Y.; Liu, C.; Shen, C.; Liu, X. Facile fabrication of durable superhydrophobic mesh via candle soot for oil-water separation. *Prog. Org. Coat.* **2019**, *136*, 105253. [CrossRef]
15. Sutar, R.S.; Latthe, S.S.; Nagappan, S.; Ha, C.; Sadasivuni, K.K.; Liu, S.; Xing, R.; Bhosale, A.K. Fabrication of robust self-cleaning superhydrophobic coating by deposition of polymer layer on candle soot surface. *J. Appl. Polym. Sci.* **2021**, *138*, 49943. [CrossRef]
16. Li, J.; Zhao, Z.; Li, D.; Tian, H.; Zha, F.; Feng, H.; Guo, L. Smart candle soot coated membranes for on-demand immiscible oil/water mixture and emulsion switchable separation. *Nanoscale* **2017**, *9*, 13610–13617. [CrossRef] [PubMed]
17. Sahoo, B.N.; Balasubramanian, K. Facile synthesis of nano cauliflower and nano broccoli like hierarchical superhydrophobic composite coating using PVDF/carbon soot particles via gelation technique. *J. Colloid Interface Sci.* **2014**, *436*, 111–121. [CrossRef]
18. Lei, T.; Xiong, J.; Huang, J.; Zheng, T.; Cai, X. Facile transformation of soot nanoparticles into nanoporous fibers via single-step electrospinning. *AIP Adv.* **2017**, *7*, 085212. [CrossRef]

19. Sutar, R.S.; Latthe, S.S.; Sargar, A.M.; Patil, C.E.; Jadhav, V.S.; Patil, A.N.; Kokate, K.K.; Bhosale, A.K.; Sadasivuni, K.K.; Mohite, S.V.; et al. Spray Deposition of PDMS/Candle Soot NPs Composite for Self-Cleaning Superhydrophobic Coating. *Macromol. Symp.* **2020**, *393*, 2000031. [CrossRef]
20. Lin, J.; Lin, F.; Liu, R.; Li, P.; Fang, S.; Ye, W.; Zhao, S. Scalable fabrication of robust superhydrophobic membranes by one-step spray-coating for gravitational water-in-oil emulsion separation. *Sep. Purif. Technol.* **2020**, *231*, 115898. [CrossRef]
21. Jaworek, A. Micro- and nanoparticle production by electrospraying. *Powder Technol.* **2007**, *176*, 18–35. [CrossRef]
22. He, T.; Jokerst, J.V. Structured micro/nano materials synthesized via electrospray: A review. *Biomater. Sci.* **2020**, *8*, 5555–5573. [CrossRef] [PubMed]
23. Kelder, E.; Nijs, O.; Schoonman, J. Low-temperature synthesis of thin films of YSZ and BaCeO$_3$ using electrostatic spray pyrolysis (ESP). *Solid State Ion.* **1994**, *68*, 5–7. [CrossRef]
24. Kourtchev, I.; Szeto, P.; O'connor, I.; Popoola, O.A.M.; Maenhaut, W.; Wenger, J.; Kalberer, M. Comparison of Heated Electrospray Ionization and Nanoelectrospray Ionization Sources Coupled to Ultra-High-Resolution Mass Spectrometry for Analysis of Highly Complex Atmospheric Aerosol Samples. *Anal. Chem.* **2020**, *92*, 8396–8403. [CrossRef] [PubMed]
25. Mirza, U.A.; Cohen, S.L.; Chait, B.T. Heat-induced conformational changes in proteins studied by electrospray ionization mass spectrometry. *Anal. Chem.* **1993**, *65*, 1–6. [CrossRef] [PubMed]
26. Surib, N.A.; MohdPaad, K. Electrospray flow rate influenced the sized of functionalized soot nanoparticles. *Asia-Pac. J. Chem. Eng.* **2020**, *15*, e2417. [CrossRef]
27. Faizal, F.; Khairunnisa, M.P.; Yokote, S.; Lenggoro, I.W. Carbonaceous nanoparticle layers prepared using candle soot by direct-and spray-based depositions. *Aerosol Air Qual. Res.* **2018**, *18*, 856–865. [CrossRef]
28. Lei, T.; Lu, D.; Xu, Z.; Xu, W.; Liu, J.; Deng, X.; Huang, J.; Xu, L.; Cai, X.; Lin, L. 2D→3D conversion of superwetting mesh: A simple but powerful strategy for effective and efficient oil/water separation. *Sep. Purif. Technol.* **2020**, *242*, 116244. [CrossRef]
29. Wang, H.; Tay, S.W.; Hong, R.S.; Pallathadka, P.K.; Hong, L. From the solvothermally treated poly (vinylidenefluoride) colloidal suspension to sticky hydrophobic coating. *Colloid Polym. Sci.* **2014**, *292*, 807–815. [CrossRef]
30. Yoon, H.; Kim, H.; Latthe, S.S.; Kim, M.; Al-Deyab, S.; Yoon, S.S. A highly transparent self-cleaning superhydrophobic surface by organosilane-coated alumina particles deposited via electrospraying. *J. Mater. Chem. A* **2015**, *3*, 11403–11410. [CrossRef]
31. Rahman, M.; Phung, T.H.; Oh, S.; Kim, S.H.; Ng, T.N.; Kwon, K. High-Efficiency Electrospray Deposition Method for Nonconductive Substrates: Applications of Superhydrophobic Coatings. *ACS Appl. Mater. Interfaces* **2021**, *13*, 18227–18236. [CrossRef] [PubMed]
32. Lei, L.; Kovacevich, D.A.; Nitzsche, M.P.; Ryu, J.; Al-Marzoki, K.; Rodriguez, G.; Klein, L.C.; Jitianu, A.; Singer, J.P. Obtaining Thickness-Limited Electrospray Deposition for 3D Coating. *ACS Appl. Mater. Interfaces* **2018**, *10*, 11175–11188. [CrossRef] [PubMed]
33. Cai, X.; Huang, J.; Lu, X.; Yang, L.; Lin, T.; Lei, T. Facile Preparation of Superhydrophobic Membrane Inspired by Chinese Traditional Hand-Stretched Noodles. *Coatings* **2021**, *11*, 228. [CrossRef]
34. Li, J.; Li, D.; Yang, Y.; Li, J.; Zha, F.; Lei, Z. A prewetting induced underwater superoleophobic or underoil (super) hydrophobic waste potato residue-coated mesh for selective efficient oil/water separation. *Green Chem.* **2016**, *18*, 541–549. [CrossRef]
35. Gao, H.; Liu, Y.; Wang, G.; Li, S.; Han, Z.; Ren, L. Switchable Wettability Surface with Chemical Stability and Antifouling Properties for Controllable Oil–Water Separation. *Langmuir* **2019**, *35*, 4498–4508. [CrossRef]
36. Cassie, A.B.D. Skyler Baxter, Wettability of porous surfaces. *Trans. Faraday Soc.* **1944**, *40*, 546–551. [CrossRef]

Review

A Comprehensive Review of Wetting Transition Mechanism on the Surfaces of Microstructures from Theory and Testing Methods

Xiao Wang [1], Cheng Fu [2], Chunlai Zhang [1], Zhengyao Qiu [1] and Bo Wang [1,*]

[1] Key Laboratory of Advanced Functional Materials, Education Ministry of China, Faculty of Materials and Manufacturing, Beijing University of Technology, Beijing 100124, China; b202009004@emails.bjut.edu.cn (X.W.); zclyouxiang@emails.bjut.edu.cn (C.Z.); qiuzhengyao@emails.bjut.edu.cn (Z.Q.)
[2] China Classification Society Quality Assurance Ltd., Beijing 100006, China; sishiergeren@126.com
* Correspondence: wangbo@bjut.edu.cn

Abstract: Superhydrophobic surfaces have been widely employed in both fundamental research and industrial applications because of their self-cleaning, waterproof, and low-adhesion qualities. Maintaining the stability of the superhydrophobic state and avoiding water infiltration into the microstructure are the basis for realizing these characteristics, while the size, shape, and distribution of the heterogeneous microstructures affect both the static contact angle and the wetting transition mechanism. Here, we review various classical models of wettability, as well as the advanced models for the corrected static contact angle for heterogeneous surfaces, including the general roughness description, fractal theory description, re-entrant geometry description, and contact line description. Subsequently, we emphasize various wetting transition mechanisms on heterogeneous surfaces. The advanced testing strategies to investigate the wetting transition behavior will also be analyzed. In the end, future research priorities on the wetting transition mechanisms of heterogeneous surfaces are highlighted.

Keywords: wetting transition; superhydrophobic; microstructures; contact angle

Citation: Wang, X.; Fu, C.; Zhang, C.; Qiu, Z.; Wang, B. A Comprehensive Review of Wetting Transition Mechanism on the Surfaces of Microstructures from Theory and Testing Methods. *Materials* 2022, 15, 4747. https://doi.org/10.3390/ma15144747

Academic Editors: Maria Vittoria Diamanti, Michele Ferrari, Massimiliano D'Arienzo and Carlo Antonini

Received: 5 May 2022
Accepted: 7 June 2022
Published: 6 July 2022

Publisher's Note: MDPI stays neutral with regard to jurisdictional claims in published maps and institutional affiliations.

Copyright: © 2022 by the authors. Licensee MDPI, Basel, Switzerland. This article is an open access article distributed under the terms and conditions of the Creative Commons Attribution (CC BY) license (https://creativecommons.org/licenses/by/4.0/).

1. Introduction

Surface wettability is one of the most vital properties of a solid surface. The wettability of a solid surface is determined by the chemical properties and the micro-texture of the surface [1–4]. Young's equation has been used to describe wetting on a smooth surface from 1805 [5]. However, real surfaces are seldom perfectly smooth. Hence, the Wenzel (W) and Cassie–Baxter (C–B) states are the two main kinds of solid–liquid wetting states on the micro-structured surfaces [6,7]. The description of the W state is based on the hypothesis that the water droplet completely penetrates the grooves of a rough surface, while the C–B state assumes the water droplet is suspended on the top of the micro-structured surface, which results in a composite interface. Compared with the W state, the C–B presents the high apparent contact angle (CA) and the low contact angle hysteresis. Maintaining the stability of the C–B state and avoiding the intrusion of water into the microstructure are essential preconditions for realizing self-cleaning, water-repelling, and anti-sticking properties [8].

The C–B state is not always stable, and the transition from the Cassie–Baxter to Wenzel state (C–B/W) can occur when it is induced by various factors, such as pressurization, [9] vibrations, [10] and the gravity of the droplet itself [11,12]. Therefore, exploring the conditions of C–B state stability and understanding the C–B/W transition mechanisms have been a central topic in the study of superhydrophobic surfaces. Over 900 journal papers studying wetting transition mechanisms have been published that cover materials science, engineering, physics, and chemistry science technology.

Here, we summarize the recent advances in the theoretical study and testing methods of the wetting state transition mechanism. In the next section, the theory of fundamental wetting models is discussed, which is followed by the description of the static contact angle model. The advanced wettability transition mechanism is presented in Section 4, and a comprehensive overview of the wetting stability testing methods is provided in Section 5. To conclude, a brief outlook for future research directions is proposed.

2. Fundamental Wetting Theory

The first investigation of wetting phenomena can be traced back to 1612, which was presented by Galileo through his report, "Bodies That Stay atop Water, or Move in it" [13]. Over the recent few decades, great progress in the wetting theories has been developed to describe the wetting state models. In this section, the fundamental wetting theories are summarized.

In 1805, Thomas Young proposed the primary law of wetting with a water droplet on a flat and smooth surface, as shown in Figure 1a [5]. Young's equation is given as:

$$\cos \theta = \frac{\gamma_{sv} - \gamma_{sl}}{\gamma_{lv}} \tag{1}$$

where θ is the static contact angle, γ_{sv}, γ_{sl}, and γ_{lv} are the solid–vapor, solid–liquid, and liquid–vapor surface tensions, respectively. Based on the value of θ, the property of the surface can be divided into the hydrophobic ($\theta > 90°$) and hydrophilic ($\theta < 90°$) surfaces [14].

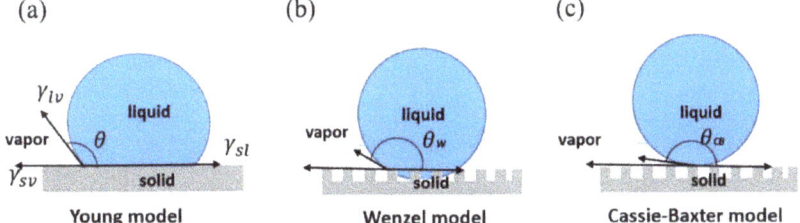

Figure 1. Various states of droplet on a solid surface. (a) Young model, (b) Wenzel model, and (c) Cassie–Baxter model.

Since Young's equation is valid only for smooth and homogenous surfaces, in 1936, Wenzel modified Young's model and introduced the roughness factor (r) to describe the wettability phenomena of the micro-structured surfaces, as shown in Figure 1b [3]:

$$\cos \theta_W = r \frac{\gamma_{sv} - \gamma_{sl}}{\gamma_{lv}} = r \cos \theta \tag{2}$$

where θ_W is the static contact angle under the Wenzel state. The roughness factor (r) is defined as the ratio of the true surface area and planar surface, which is higher than 1 for a microstructured surface. In 1945, Cassie–Baxter described another wetting state for the droplet on microstructured surface, shown in Figure 1c. The model supposed the droplet is suspended on the top of the micro-structured surface, which results in a composite interface [4]. In the C–B model, the apparent contact angle is influenced by the contribution of two different phases, as described in the equation below:

$$\cos \theta_{CB} = f_{sl} \cos \theta_{sl} + f_{la} \cos \theta_{la}, \tag{3}$$

where the θ_{CB} is the static contact angle under the Cassie–Baxter state, f_{sl} and f_{la} represent the surface fractions of the phases of solid–liquid and liquid–air, and θ_{sl} and θ_{la} represent the corresponding contact angles. Since the θ_{la} is $180°$ in C–B state, so $\cos \theta_{la} = \cos 180° = -1$, then Equation (3) can be rewritten as:

$$\cos \theta_{CB} = f_{sl} \cos \theta_{sl} - f_{la} \quad (4)$$

Lafuma et al. [9] derived the C–B/W transition in 2003. As shown in Figure 2, where θ^* is the apparent contact angle when minimizing the surface energy of a drop on a rough substrate, θ is the apparent contact angle under Young's state, $\varnothing_s$ is the fraction of solid in contact with the liquid, and θ_c is denoted as the critical contact angle between the two states. The coordinates of A, B, C, and D are ($\cos 180°$, $\cos 180°$), ($\cos \theta_c$, $\cos \theta^*$), ($\cos 90°$, $\cos 90°$), and ($\cos 90°$, $\cos \theta^*$), respectively. When the apparent contact angle θ is larger than θ_c, it follows the C–B state model that the air grooves would be trapped below the drop to form the composite contact. When $90° < \theta < \theta_c$, the two states might coexist. Various studies have validated the existence of a metastable state based on simulation and experiment results [15–17].

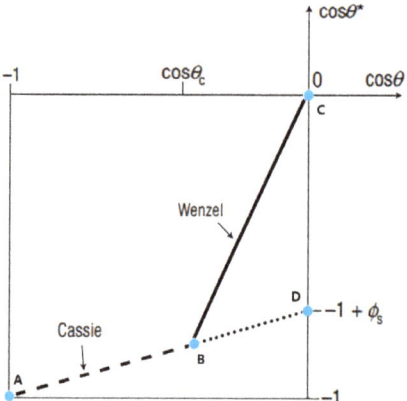

Figure 2. Two models of superhydrophobicity. The solid line and dashed line represent Wenzel state and Cassie–Baxter state, respectively. The dotted line represents the metastable situation where the C–B state can also be observed for $\theta < \theta_c$. The coordinates of A, B, C, and D are ($\cos 180°$, $\cos 180°$), ($\cos \theta_c$, $\cos \theta^*$), ($\cos 90°$, $\cos 90°$), and ($\cos 90°$, $\cos \theta^*$), respectively. Reprinted with permission from Ref. [9]. Copyright@2003, Nature Publishing Group.

In recent decades, with the development of computer science, computer simulation has become a vital research method. There are numerous methods that can be used to simulate the droplet states on superhydrophobic surface. From the microscale state, molecular dynamics (MD) study the flow behavior of statistical fluid from the perspective of molecular atoms to explore the wetting state transition of droplets at molecular scale [18–28]. The simulation dynamics are sometimes given by the Monte Carlo method. Bryk et al. [29] (2021) found that the effective interface potential method can be used to determine the location of the critical wetting transition. Lopes et al. [30] (2017) presented a Potts model simulation of the C–B/W transition on a surface decorated by a regular distribution of pillars.

Meanwhile, for a larger scopic state, computational fluid dynamics (CFD) are normally widely used. They include the lattice Boltzmann method (LBM) and the finite volume method (FVM). LBM is a good choice from mesoscopic state for the simulation of superhydrophobic surface. There are numerous studies of superhydrophobic surface simulation using LBM, such as contact angle of microdroplets on superhydrophobic surface [31–36] and wetting transition between Cassie and Wenzel [37,38]. FVM is a mature algorithm in the field of CFD. It is often used to solve macroscopic state problem instead of wetting state transition.

3. Corrections in the Static Contact Angle Model on Heterogeneous Surface

The W and C–B models measure the static contact angle with a solid phase area fraction over the whole surface, but the assumption is not necessarily satisfied for heterogeneous surfaces with complex morphologies [39]. McHale [40] (2007) indicated that the roughness ratio of the W model and frictional contact area are valid only when the surface is isotropic all over with uniform morphology. Therefore, it is very important to build a more accurate and versatile description system to precisely reflect the relationship between the microstructure criteria and static contact angle. At present, the surface microstructure description system includes a roughness description system, fractal theory description system, re-entrant geometry description system, and contact line description system. Next, we will review the advanced progress in these description systems.

3.1. Roughness Description System

The first statistical description of surface roughness can be traced back to 1966 evolved from tribology. Greenwood and Williamson proposed a theory based on the assumption that the height of the rough surface contour obeys a Gaussian distribution, shown in Figure 3a [41]. However, the detailed textures of a rough surface of a hydrophobic material could elegantly affect its wetting performance. Jiang, et al. [42] (2020) summarized the three typical types of structural morphologies that can change the surface wetting properties: pillar-structured surfaces, pore-structured surfaces, and groove-structured surfaces. Kim et al. [43] (2020) demonstrated that the arrangement of the pattern also had a great correlation with the static contact angle by experiment. They investigated the four shapes of pattern arrangement (triangular, square, hexagonal, and octagonal). Cao et al. [44] (2021) found a correlation between the depth-to-width ratio and static contact angle with the same surface morphology by experiment. As the depth-to-width ratio increases, the air–solid contact area also increases, which leads to an increase in the contact angle.

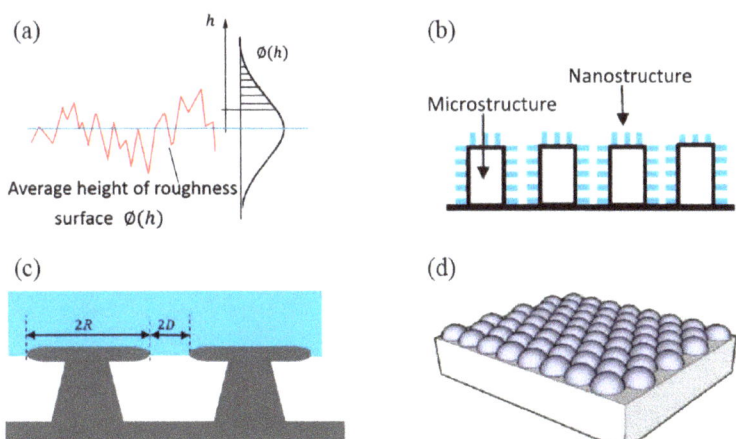

Figure 3. The schematic diagram of microstructures model. (**a**) Roughness model: assuming that the height of rough surface profile obeys Gaussian distribution. Reprinted with permission from Ref. [41]. Copyright@1966, Proc. R. Soc. (**b**) Fractal geometry model. Reprinted with permission from Ref. [45]. Copyright@1967, Science (New York, NY, USA). (**c**) Re-entrant structure model. Reprinted with permission from Ref. [39]. Copyright@2020, Journal of Physical Chemistry. (**d**) Typical nanoparticle coating model. Reprinted with permission from Ref. [46]. Copyright@2002, Nano Lett.

3.2. Fractal Theory Description System

When building a microstructure surface model, dimension scale plays an important role in analyzing the influence of surface microscopic characteristics on the droplet wetting state. There are many scales related to wetting, such as the atomic scale ($10^{-10} \sim 10^{-9}$ m), microscopic scale ($10^{-9} \sim 10^{-6}$ m), mesoscopic scale ($10^{-6} \sim 10^{-2}$ m), and macroscopic scale (>10^{-2} m) [47]. However, since the modeling parameters in the statistical model are related to the dimension scale of the rough surface, which is affected by the resolution of the measuring instrument and the sample length of the rough surface, they are important for accurate modeling. Hence, the fractal geometry, which combined the different scales as shown in Figure 3b, is introduced for the analysis of the rough surface contact angle problem [45]. The multi-scaled wetting contact angle of θ_{CB} and θ_W can be expressed by Equations (5) and (6), based on the fractal theory:

$$\cos\theta_{CB} = f\left(\frac{L}{l}\right)^{D-2}\cos\theta - f_{la} \tag{5}$$

$$\cos\theta_W = \left(\frac{L}{l}\right)^{D-2}\cos\theta \tag{6}$$

where D is the Hausdorff dimension, i.e., $D = \log(4)/\log(3) = 1.2618$, and $\left(\frac{L}{l}\right)^{D-2}$ is the surface area magnification factor [48]. L and l are the upper and lower limit scales of the fractal structure surface, and θ is the intrinsic contact angle of the material.

Both random and ordered fractal structures have positive influences on the superhydrophobic performance, as the fractal theory illustrated [49,50]. Meanwhile, limited to the self-similarity and self-affinity, the complexity of micro/nanostructures is not satisfactorily described by the fractal theory [45]. Davis et al. [51] (2017) designed three fractal structures, but the results showed no clear correlation between the static contact angle and the fractal dimensions.

3.3. Re-Entrant Geometry Description System

Tuteja et al. [52] (2007) demonstrated that there is a third description system related to the wetting performance, called "re-entrant geometry". It exhibited the capability of supporting repelling behavior to a droplet [53,54]. The re-entrant structures have distinctive features with a wider top and a narrower bottom, shown in Figure 3c. The typical re-entrant geometry can be realized with various possible geometries, such as micro-mushrooms, micro-hoodoo arrays, fiber mats, micro-nail forests, micro-posts, and nanoparticle coatings [39]. The contact angle of micro-hoodoo arrays can be described with the spacing ratio by the following equation:

$$\cos\theta_{CB} = -1 + \frac{1}{D^*}[\sin\theta + (\pi - \theta)\cos\theta] \tag{7}$$

where D^* is the spacing ratio, expressed as $D^* = (R+D)/R$, in which R is the radius of the micro-hoodoo, D is the half distance between micro-hoodoos. θ is the intrinsic contact angle of the material. For a typical coated nanoparticle model, shown in Figure 3d, the θ_W and θ_{CB} can be expressed by Equations (8) and (9), respectively [55]:

$$\cos\theta_w = \left[1 + \frac{\pi}{4\sin\alpha}\left(\frac{2R}{D}\right)^2\right]\cos\theta \tag{8}$$

$$\cos\theta_{CB} = \frac{\pi}{4\sin\alpha}\left(\frac{2R}{D}\right)^2 - 1 \tag{9}$$

where R is the radius of the nanoparticle. D is the distance between the centers of two adjacent nanoparticles. α is the angle of the diamond cell. θ is the intrinsic contact angle of the material.

3.4. Fractal Theory Description System

Both the Wenzel and the Cassie–Baxter theories calculate the apparent contact angle from the solid–liquid contact area. The actual measurement results of the contact angle may not be consistent with the two classical theoretical models [56,57]. The solid–liquid contact results from a microscopic surface demonstrated a significant impact of the contact line on the surface wettability. Extrand et al. [58] (2002) proposed that the solid surface energy and the microstructure of the contact line, rather than the inside geometry of the contact area, were the main factors affecting the apparent contact angle. Gao et al. [57] (2007) prepared three groups of microstructures with different morphologies and different surface energies. It was found that the microstructure below the droplet did not affect the contact angle. Oner et al. [59] (2000) found that, when the solid–liquid area fraction was constant, the contact angle would increase as a result of the decreases in the contact length of the three-phase contact line.

4. Corrections in the Wetting Transition Mechanism on Heterogeneous Surface

During the last two decades, there has been a drastic upsurge in the research publications related to the correction mechanisms of the C–B/W transition on a heterogeneous surface [60–64]. In this section, the corrections of the wetting transition mechanism are classified from three representative microstructure surfaces: flat-top pillar microstructure, multi-scale microstructure, and re-entrant microstructure.

4.1. The Universal Transition Mechanism on Flat-Top Pillar Microstructure

Since the flat-top pillar is the simplest rough structure, it was used as a representative model in the study of wetting transition [65,66]. For the longitudinal propagation of the liquid, there are two ways in which transition can be induced. The first way is a depinning mechanism in which the interface is curved due to the Laplace pressure inside the droplet, shown in Figure 4a. When the hanging interface cannot remain pinned at the pillar edges, the second way of the sag mechanism is induced with the curved liquid–air interface touching the bottom, as shown in Figure 4b [11,65–71], where θ_e is the intrinsic contact angle of microstructure sidewall, θ_{pin} is the contact angle of liquid–gas interface pinned at the sidewall of the microstructure, sag is the sag depth at the top of the curved interface, and H is the depth of the microstructure. For the lateral propagation of the liquid, Ren et al. [72] (2014) found that the propagation of the liquid front proceeded in a stepwise manner by numerical simulation, shown in Figure 4c,d. Lateral propagation of the liquid front proceeds by one layer of the grooves, from W_1 to W_3. W_2 is an intermediate metastable state (a local minima). S_2 is the transition state (saddle point) from W_1 to W_2, and S_3 is the transition state from W_2 to W_3.

For these flat-top pillar microstructures, various models have been proposed to explain the transition mechanisms and criteria of the C–B/W transition, which can be mainly divided into the thermodynamic analysis and the force-based analysis. Thermodynamic analysis minimizes the Gibbs energy of the system [65,66,73,74], while the force-based analysis establishes the balance of the capillary forces near the three-phase contact line (TPCL) [68,75–77].

From the thermodynamic analysis perspective, since the C–B state has a higher energy state than W state, the droplet penetration to the grooves is accompanied by a decrease in the Gibbs energy. This is formed with two components. One is due to the replacement of the solid–air interface with the solid–liquid interface, and the other is due to the change in the liquid–air interfacial area [78–81]. Although the C–B/W wetting transition is energetically favored, Patankar et al. [65] (2004) conducted a theoretical study to present an energy barrier between the two states on a pillar-patterned surface, which requires extra work to drive the transition with limited kinetics. The barrier energy for the C–B/W transition is given as Equation (10) [65]:

$$G_{B1} = G_c - (r-1)\cos\theta_e \sigma_{lv} A_c \tag{10}$$

with G_c defined in Equation (11):

$$G_c = S_c\sigma_{lv} - \cos\theta_r^c A_c \tag{11}$$

where G_c is the energy of a Cassie droplet on a rough substrate, G_{B1} is the barrier energy for the transition of a Cassie droplet to a Wenzel droplet. θ_e is the equilibrium contact angle of the liquid droplet on the flat surface. θ_r^c is the apparent contact angle of a drop under Cassie state. S_c is the area of the liquid–vapor contact for a Cassie droplet. A_c is the area of contact with the substrate projected on the horizontal plane under the Cassie state. σ_{lv} is the liquid–vapor surface energy per unit area. The barrier energy of the C–B/W transition by considering the sag state is given as Equation (12):

$$G_{B2} = G_w + (1 - \varnothing_s)(1 + \cos\theta_e)\sigma_{lv}A_W \tag{12}$$

with G_w defined in Equation (13):

$$G_w = S_w\sigma_{lv} - \cos\theta_r^w A_w \tag{13}$$

where G_{B2} is the barrier energy for the Wenzel droplet without forming the liquid–solid contact at the bottom of the valleys. G_w is the energy eventually reaching the equilibrium shape of a Wenzel droplet. $\varnothing_s$ is the area fraction on the horizontal projected plane of the liquid–solid contact. S_w is the liquid–vapor surface area of a Wenzel droplet. A_w is the area of contact with the substrate projected on the horizontal plane under the Wenzel state.

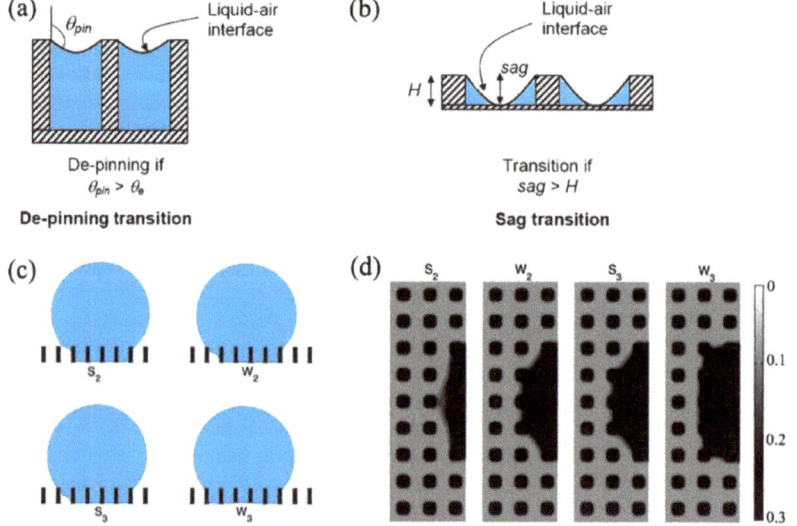

Figure 4. The schematic diagram of the droplet longitudinal propagation in flat-top pillar microstructure for (**a**) depinning transition, and (**b**) sag transition. Reprinted with permission from Ref. [66]. Copyright@2004, Langmuir. (**c**) The schematic diagram of droplet lateral propagation, and (**d**) the simulation results. Reprinted with permission from Ref. [72]. Copyright@2014, Langmuir.

From the force-based equilibrium perspective, the resistance to the liquid penetration was considered to be the force produced by the liquid–gas interfacial tension acting on the protrusion side surfaces through TPCL. By considering the force equilibrium between both capillary forces, the influence of gravity and the external pressure, different expressions of the critical pressure were derived to study the C–B/W transition on a micro-structured surface. Xue et al. (2012) gave a theoretical model for predicting the critical pressure on the submersed substrates formed with the cavities and pillars [15]. It was found that both pillars' and cavities' geometries existed in the metastable state after depinning. The theory had good agreement with the experiment by Lei et al. (2010), which demonstrated the characteristic size of pillars and that the solid fraction played more important roles than the pillar's arrangement on the hydrophobicity with higher critical pressure [82]. Zheng et al. (2005) gave the universal critical pressure (p_c) formulation of pillars as below [75]:

$$p_c = -\frac{\gamma f \cos \varnothing_0}{(1-f)\lambda} \quad (14)$$

where λ is the pillar slenderness ratio, f is the fraction of the projection area that is wet, and $\gamma f \cos \varnothing_0$ is the water–air interfacial tension.

4.2. The Asymmetric Wetting Propagation

Both lateral and longitudinal propagations were investigated for the asymmetric wetting propagation [83–85]. For the lateral propagation, Fetzer et al. [86] (2011) conducted experimental work to explain that the lateral asymmetries can be attributed to the curvature of the contact line and the different mechanisms of depinning, such as nucleated jump-like motion and continuous depinning from the sides. Priest et al. [83] (2009) found the asymmetries were attributed to the continuity of the solid component by experiment, and this behavior was consistent with the wettability of chemically heterogeneous surfaces.

For the longitudinal propagation, when the liquid–gas interface touches the bottom of the microstructure, there are two possible contact modes for the sag mechanism: symmetric contact and asymmetric contact [73,85,87,88]. The asymmetric contact shortens the progression of the metastable state to the Wenzel state; hence, it may affect the lifespan of superhydrophobicity [85]. Kim et al. [89] (2018) used a numerical method to find there is an asymmetric depinned stage during the wetting transition process, shown in Figure 5. The wetting transition of a cylindrical cavity begins with an axially symmetrically pinned interface of the liquid and vapor. It is followed by a symmetric depinned interface and then the formation of an annular interface. Finally, the asymmetric depinned interface was formed before reaching the Wenzel state. Giacomello et al. [88] (2012) explained the reason for the asymmetric using the free energy minimization theory. At low filling levels, the interface with the minimized free energy is a straight line, while, for higher liquid volumes in the box, a quarter of the circle occupying one corner offered the minimal free energy.

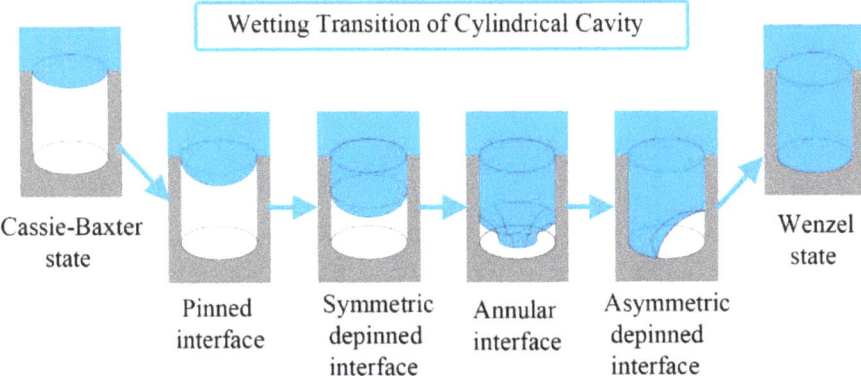

Figure 5. The asymmetric wetting transition mechanism of flat-top pillar microstructure of cylindrical cavity. Reprinted with permission from Ref. [89]. Copyright@2018, the Journal of Physical Chemistry.

4.3. The Wetting Transition Mechanism on Multi-Scaled Microstructure

Various biomimetic studies found that multi-scaled microstructures can enhance the hydrophobicity of natural surfaces, with a typical example of a lotus leaf [73,90–95]. It mainly includes two ways: 1. the droplet infiltrates the nanostructures, 2. multi-scaled microstructure provides more pinning points during the depinning stage. Huang et al. [8] (2013) and Bormashenko et al. [96] (2015) found there is a typical stage that the droplet suspended on the microstructure can infiltrate the nanostructures under lower pressure. Meanwhile, Hemeda et al. [97] (2014) and Xue et al. [15] (2012) discovered that the multi-dimension of microstructures provided more pinning points for the liquid–gas interface during the C–B/W transition. Many studies have investigated these wetting transition mechanisms with different methods. Zhang et al. [98] (2013) and Lee et al. [99] (2016) used the lattice Boltzmann method to investigate the C–B/W wetting transition on the multi-scaled microstructures. Shen et al. [100] (2015) used an experimental method to investigate the wetting transition mechanism on a Ti_6Al_4V micro-nanoscale hierarchical structured hydrophobic surface. They demonstrated that the wetting transition process not only increased the apparent contact angle but also decreased the sliding angle significantly. Teisala et al. [101] (2012) used the experimental method to generate a hierarchically rough superhydrophobic TiO_2 nanoparticle surface by the liquid flame spray. It was found that a wetting transition occurred on a superhydrophobic surface at the nanometer scale.

The energy models were also used to explain the reason for the transition mechanisms on the multi-scaled surface. Gao et al. [57] (2006) explained that the micro/nanostructure makes C–B state wetting energetically favorable. The additional small-scale roughness on the side surface of the hydrophobic pillars increased the potential barrier for the C–B/W transition, thus making the C–B wetting state more stable. Liang et al. [102] (2017) built a 3-D model to analyze the wetting behavior from a thermodynamics perspective, shown in Figure 6. It shows the variations in normalized free energy (NFE) with apparent contact angle for C–B, C–B metastable, and W states. Here, the NFE decreases at first and then increases with the increase in the contact angle. However, the NFE curve of the C–B state is lower than that of the other two states. Nosonovsky et al. [103] (2009), Hejazi et al. [104] (2013), and Huang et al. [8] (2013) explained this with the capillary mechanisms by both computational and experimental work. They reported that the microstructures or defects on the substrates can significantly increase the wetting hysteresis due to three-phase contact line (TPCL) pinning. As illustrated in Figure 7, the bumps may pin the liquid–air interface because an advance in the liquid–air interface could result in a decrease in the contact angle, which provides the stability of the composite interface.

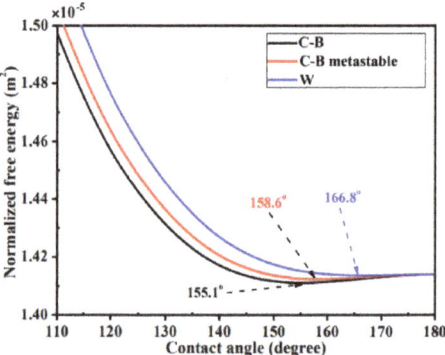

Figure 6. Schematic illustration for the variations in normalized free energy with apparent contact angle for C–B, C–B metastable, and W states on multi-scaled microstructure. Reprinted with permission from Ref. [102]. Copyright@2017, Physicochemical and Engineering.

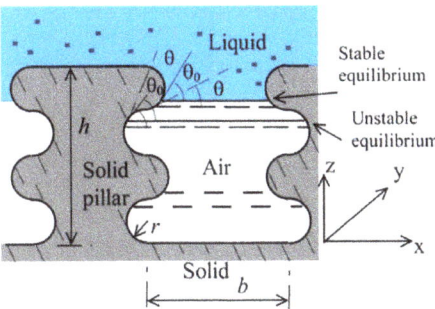

Figure 7. Stability analysis of the composite interface consisting of a rough surface with two-dimensional pillars with semi-circular grooves. Reprinted with permission from Ref. [103]. Copyright@2009, Elsevier Ltd.

4.4. The Wetting Transition Mechanism on Re-Entrant Microstructure

Cai et al. [105] (2019) studied three types of nanostructures with different longitudinal-section geometries, including base angles of 60° (inverted trapezoid), 90° (rectangular), and 120° (regular trapezoid). It was shown that the inverted trapezoidal nano-structure surface helped to keep the droplet in the C–B state, in which liquid did not penetrate the nano-structure. This was also described in Refs. [106,107]. Savoy et al. [108] (2012) used a molecular dynamics method to simulate the wetting behavior of different-size droplets on a "T" shape structure via boxed molecular dynamics, which is a technique that is used to quantify the free-energy landscape and estimate the transition rate as the drop moves from one low free-energy basin to another. Further, they found that, at the same height, the "T" structure surface needs to overcome a higher energy barrier than that of the square column surface, which somehow enhanced surface evacuation (shown in Figure 8). Wang et al. [109] (2019) validated the reason for the high superhydrophobicity of a "T" structure by both experimental work and a simulation method. The strong pinning effect on the contact line can significantly change the contact angle and wetting state of droplets.

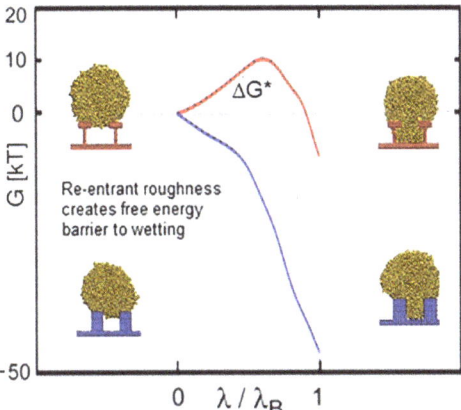

Figure 8. Re-entrant roughness creates a higher energy barrier to wetting. Reprinted with permission from Ref. [108]. Copyright@2012, Langmuir.

5. Wetting Transition Testing Methods

Lafuma et al. [9] (2003) first proposed a testing method by squeezing droplets with two superhydrophobic surfaces to reflect the wetting stability. This method can be used to observe and study the droplet's critical pressure and contact angle. The experiment mainly utilizes a micro force sensor on an optical microscope platform. However, this test is only valid for some superhydrophobic materials with poor wetting stability. In recent years, several new versatile strategies were proposed to investigate the wetting transition behavior. The methods include (1) optical, (2) acoustic, (3) confocal laser scanning microscopy, (4) freeze stripping, and (5) high-speed camera methods. The principles and merits of these measurement methods are listed in Table 1, and they will be explained in detail in the following subsections.

Table 1. Summary of wetting transition testing methodology.

Method	Principle	Merits	Reference
Optical reflection	Different wetting states show different light reflection intensity	Simplest and direct	[8,110–114]
Optical diffraction	Change in diffraction pattern reflects the change in gas layer thickness	The shape of the liquid–gas interface can be calculated	[82,115]
Confocal laser scanning microscopy	Scanning the samples by fault section, and three-dimensional reconstruction	Real-time observation of wetting state transition process	[16,61,116–118]
High-speed camera	Very short exposure time	High temporal resolution	[11,118–120]
Freeze fracture	a certain interface was immersed in liquid nitrogen, and the droplet is frozen rapidly	Small applicable scale for nano-scale microstructure surface	[121–123]
Acoustic	The differences of reflection of longitudinal acoustic waves at the composite interface	Versatile and integrable	[106,124–126]

5.1. Optical Methodology

The optical methodology includes optical reflection and diffraction methodologies. The reflection methodology is the simplest and the most direct way [8,110–114]. It measures the total reflection from an underwater superhydrophobic interface to investigate the wetting behavior and critical pressure of the C–B/W transition. The superhydrophobic interface has a unique reflection property underwater. At the Cassie–Baxter state, the gas–liquid interface satisfies the light total reflection conditions, and its reflected light is relatively bright, shown in Figure 9a. However, at the Wenzel state, since the liquid has penetrated the voids of the microstructure, there is only reflection light at a rough interface, which is darker than that at the Cassie–Baxter state, shown in Figure 9b. Meanwhile, in the meta state, the intensity of the reflected light is between those from the two states. Huang et al. [8] (2013) used reflection methodology to investigate the wetting behavior and to measure the critical pressure of C–B/W transition. Since it is difficult to effectively quantify the intensity of the reflected light by visual observation, laser illumination and photoelectric detection were introduced to quantify the intensity of the reflected light as a function of the reflected light under pressure, as shown in Figure 9c. The critical pressure of the C–B/W transition can be obtained from the inflection point on the reflection curve.

Compared to the reflection methodology, the optical diffraction methodology can establish the shape of the liquid–gas interface. When a superhydrophobic surface is submerged in water in a fluidic chamber, the surface pattern consisting of regular pillars diffraction can be observed with a laser beam passing through the submerged grating sample in water. The shape of the liquid–gas interface can be indicated by measuring the intensity of several diffraction spots. Lei et al. [82] (2010) employed this method to control and monitor the switching of the C–B/W transition. As shown in Figure 10, water was injected into the chamber with a syringe through the inlet and outlet system. After blocking the outlet valve, hydraulic pressure can be applied through the inlet. A laser beam was aligned to pass through the fluidic chamber and a charge-coupled device (CCD) camera was used to capture diffraction images as a function of the applied pressure. The pressure for the transition between two states can be quantitatively measured.

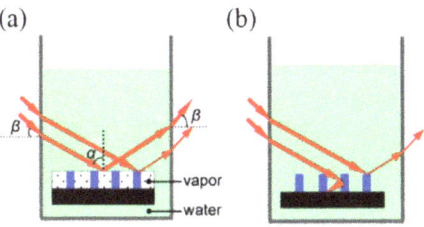
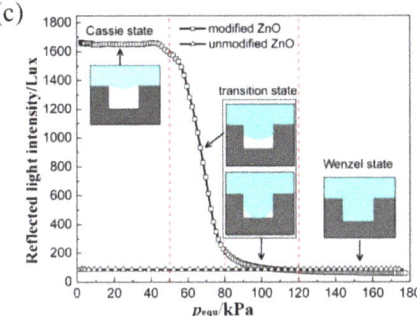

Figure 9. Schematic illustration for the optical reflection methodology: (**a**) the reflection characteristic of Cassie state, (**b**) the reflection characteristic of Wenzel state, and (**c**) the intensity of reflection light vs. equivalent pressure. Reprinted with permission from Ref. [8]. Copyright@2013, Acta Phys. -Chim. Sin.

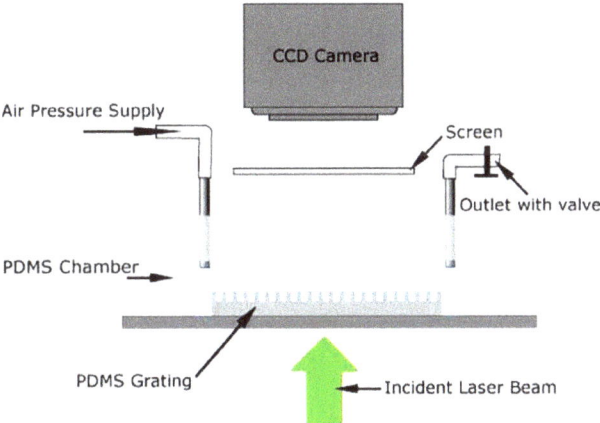

Figure 10. Schematic diagram of the experimental setup for the pressure-dependent observation of diffraction patterns of the water-submerged superhydrophobic grating. Reprinted with permission from Ref. [82]. Copyright@2010, Langmuir.

Through the optical diffraction methodology, Rathgen et al. [115] (2010) studied the microscopic shape, contact angle, and the Laplace law behavior at the liquid–gas interfaces on a superhydrophobic surface.

5.2. Acoustic Methodology

As the optical methods are neither versatile nor integrable, Dufour et al. [106] (2013) presented an alternative method based on acoustic measurements. An acoustic transducer is integrated on the backside of a superhydrophobic silicon surface with a droplet deposited on the superhydrophobic surface. By analyzing the reflection of longitudinal acoustic waves at the liquid–solid–vapor interface, the transition of C–B/W can be tracked by measuring the reflection coefficient, shown in Figure 11. Here the Pillar dimensions are diameter a = 15 µm, pitch b = 30 µm, and height h = 20 µm, and the thickness of bulk silicon is $e_{Si} \approx 400$ µm. The $r_{top}{}^*$ and $r_{bottom}{}^*$ are the normalized acoustic reflection coefficients at the top and bottom part of a micropillars array, respectively. With a plane acoustic wave propagating onto a micro-structured superhydrophobic surface with two different acoustic media, the absolute value of the reflection coefficient $R_{2/1}$ can be calculated from Equation (15):

$$|R_{2/1}| = (\rho_2 v_2 - \rho_1 v_1)/(\rho_2 v_2 + \rho_1 v_1) \tag{15}$$

where ρ is the density of the medium and v is the velocity of the acoustic wave. They also measured the evolution of reflection coefficients on the top and bottom parts of the pillars during the changing of the droplet. The results are shown in Figure 11b. At time = 0 min, an acoustic measurement was performed without a droplet, so the reflection coefficients at the bottom and top interface equal 1. By applying a droplet on the surface at 2 min, the reflection coefficient was reduced to 0.83, while the reflection coefficient of the bottom was not affected. Evaporation occurred after 2 min, and to 11 min. Since the drop was in a meta-Cassie wetting state, no notable change was observed for the reflection coefficients during this time. A spontaneous wetting transition was observed after 11 min with a sudden decrease in the reflection coefficients for both the top and bottom interface. Finally, the drop was evaporated, with the recovery of the coefficients to 1.00. The traces of the coefficients effectively described the wetting transition process. The acoustic methodology was also used to study the wetting transition process with the change in the droplet densities [124].

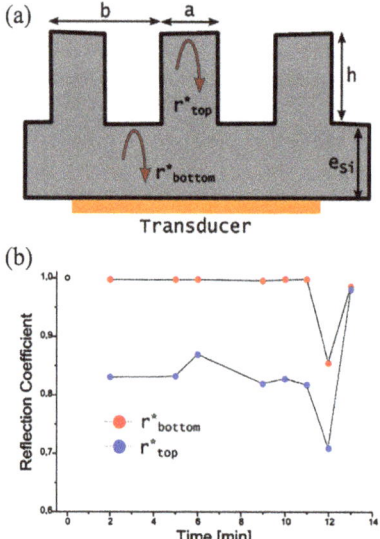

Figure 11. (a) Schematic diagram of the experimental setup for the acoustic methodology. (b) The evolution of reflection coefficients on the top and bottom interface with the droplet evaporation. Reprinted with permission from Ref. [106]. Copyright@2013, Langmuir.

5.3. Confocal Laser Scanning Microscopy Methodology

Confocal laser scanning microscopy (CLSM) can continuously record the reflected light during the wetting state transmission through a direct 3D nondestructive imaging method. The setup is shown in Figure 12. The conventional 2D optical observation provides only limited and no semi-quantitative information about the topological complex at the water–gas interface. Since the interface below the liquid cannot be imaged by using a scanning electron microscope (SEM) or transmission electron microscope (TEM), CLSM could provide a better measurement than an SEM, TEM, or atomic force microscope (AFM) [116]. Luo et al. [116] (2010) used a CLSM to observe the air trapped in the buried interface. They presented two approaches to control the wetting state transition by either ultrasonic treatment or introduction of a surface wetting agent, such as sodium dodecylbenzene sulfonate (SDS), into the droplet. Papadopoulos et al. [61] (2013) imaged the dynamic collapse of the C–B state process in detail. They presented the asymmetric contact of the water–gas interface under the metastable evolution process using a CLSM with five detectors.

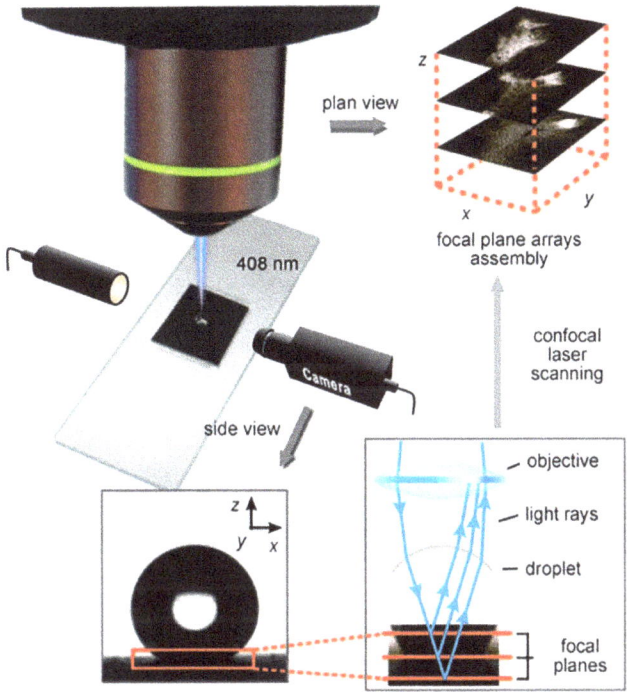

Figure 12. Schematic illustration for the CLSM methodology. Side view and the CLSM 3D plan view for observing a droplet on a superhydrophobic surface sheet. Reprinted with permission from Ref. [116]. Copyright@2010, Wiley-VCH Verlag GmbH & Co. KGaA, Weinheim.

5.4. Freezing Fracture Methodology

The freezing fracture methodology can be applied for direct observation of the wetting transition process of a droplet on a multi-scaled microstructure [123]. The droplet system at a certain wetting state is immersed in liquid nitrogen, and the droplet would be frozen immediately. The frozen droplet is observed by a scanning electron microscope. The set-up of the experiment is shown in Figure 13. First of all, a liquid droplet of deionized water is deposited into a holder, shown in Figure 13a, a nano-patterned surface of a Si wafer is pressed onto the holder, followed by rapid freezing at 77 K, shown in Figure 13b. Next, the patterned Si wafer is detached from the holder, shown in Figure 13c, which is placed in an evaporation chamber equipped with a cooling stage. Layers of 3 nm Pt and 5 nm C are deposited by electron beam evaporation onto the fracture to avoid sublimation. Ensikat, et al. [127] (2009) applied this method to visualize the contact area between liquids and superhydrophobic biological surfaces. To avoid the sublimation of the droplet and to stabilize the imprint, Cannon et al. [121] (2010) modified the standard method by coating the replica surface with thin platinum and carbon layers.

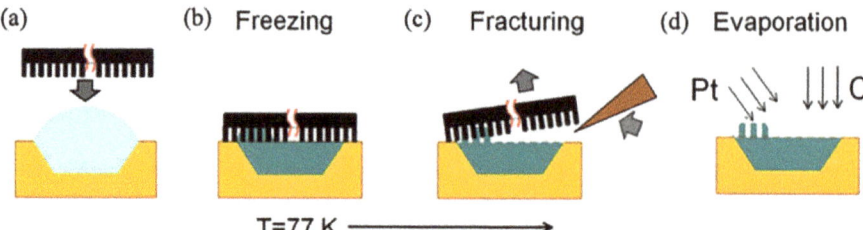

Figure 13. Schematic illustration for the freezing fracture methodology. (**a**) Before freezing, (**b**) rapid freezing of the pure water, (**c**) fracturing, (**d**) surface coated by thin platinum and carbon layers. Reprinted with permission from Ref. [123]. Copyright@2013, Langmuir.

5.5. High-Speed Camera Methodology

The high-speed camera can visualize the bouncing of the droplets on a microstructure surface. Hao et al. [120] (2015) investigated the droplet impact dynamics by reflection interference contrast microscopy (RICM) with a wavelength of 546 nm. The process was also recorded by a high-speed camera with a frame rate of 50,000 fps. Li et al. (2010) also used this method to capture the dynamic behavior of the droplet on different surfaces. As shown in Figure 14a–e, there were five textured surfaces T_{10}^{20}, T_{10}^{40}, T_{20}^{40}, T_{20}^{80}, and T_{20}^{100}, respectively, where the textured surfaces are signified by the T_D^P. D is the diameter of the pillars, and P is the distance between to pillars. The image on each patterned surface was very similar except for the counterparts on the T_{10}^{20} and T_{20}^{40} surfaces [128].

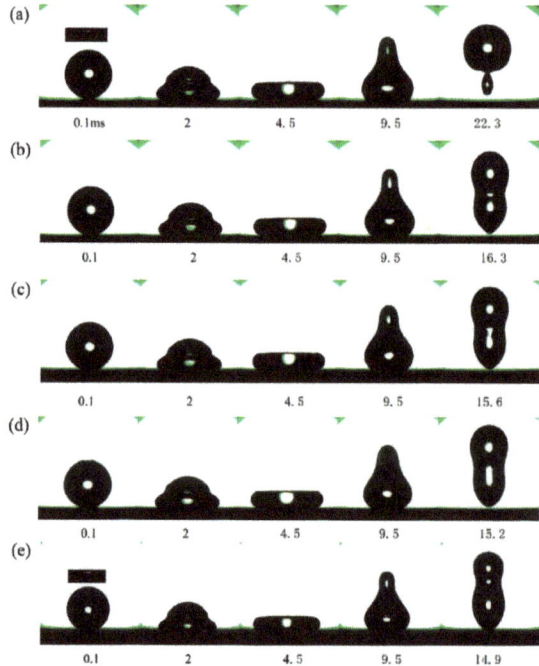

Figure 14. Schematic illustration for the result of high-speed camera methodology. Sequential images show the temporal evolution of the dynamic impact drop on different surfaces. The impact velocity was set up to 0.54 m/s, and then parts (**a–e**) represent drop events on the T_{10}^{20}, T_{10}^{40}, T_{20}^{40}, T_{20}^{80}, and T_{20}^{100} textured surfaces, respectively. Reprinted with permission from Ref. [128]. Copyright@2010, Langmuir.

6. Conclusions and Future Outlook

In this review, a general outline of the fundamental theories on wettability was discussed first, followed by the illustration of recent developments in the static contact angle models for different heterogeneous surfaces. Different description systems were also discussed, including general roughness description, fractal theory description, re-entrant geometry description, and contact line description. Further, the influence of different microstructures on the transition mechanism from the Cassie–Baxter regime to the Wenzel regime has been discussed. The knowledge regarding the available experimental approaches is critical in guiding the selection for different purposes and applications. Therefore, the different measurement approaches are summarized in this review, including optical methodology, acoustic methodology, confocal laser scanning microscopy methodology, freeze stripping methodology, and high-speed camera methodology. A broad outlook for potential future research on the wetting transition mechanism is listed as follows:

- At present, the metastable state is mainly described based on the energy barrier and Laplace pressure, with certain limitations. Most of the theories for the metastable state are based on regular periodic arranged structures. Hence, only the materials with a regular microstructure can have their superhydrophobic properties predicted approximately. Therefore, a thorough understanding of the theory of energy barrier and Laplace pressure on different heterogeneous surfaces is essential.
- There is also an asymmetric contact configuration on the microstructure surface when wetting is metastable. At present, there is a lack of a distinct calculation model for asymmetric instability from a theoretical perspective.
- Most of the classical models on the transition mechanism with fractural geometry and re-entrant geometry assume the droplet perpendicularly impacting the surface. However, inclined surfaces are more common in reality. Hence, future research should focus on droplet dynamics over inclined surfaces.
- The new testing techniques are essential to further discern and identify underlying issues in wetting study since the wetting transition of the superhydrophobic state under pressure is a complicated process. Therefore, future study on advanced testing methods is necessary.

Author Contributions: Conceptualization, X.W., C.F. and B.W.; validation, X.W., C.Z. and Z.Q.; investigation, X.W. and C.F.; writing—original draft preparation, X.W.; writing—review and editing, B.W.; visualization, X.W. and Z.Q.; supervision, B.W. All authors have read and agreed to the published version of the manuscript.

Funding: This research received no external funding.

Institutional Review Board Statement: Not applicable.

Informed Consent Statement: Not applicable.

Data Availability Statement: Not applicable.

Conflicts of Interest: The authors declare no conflict of interest.

References

1. Iskender, Y.; Yilgör, E.; Koşak, Ç.S. *Superhydrophobic Polymer Surfaces: Preparation, Properties and Applications*; With Stimuli-Responsive Surfaces with tunable wettability; Smithers Rapra: Shropshire, UK, 2016.
2. Erbil, H.Y.; Cansoy, C.E. Range of Applicability of the Wenzel and Cassie–Baxter Equations for Superhydrophobic Surfaces. *Langmuir* **2009**, *25*, 14135–14145. [CrossRef] [PubMed]
3. Cansoy, C.E.; Erbil, H.Y.; Akar, O.; Akin, T. Effect of pattern size and geometry on the use of Cassie–Baxter equation for superhydrophobic surfaces. *Colloids Surf. A Physicochem. Eng. Asp.* **2011**, *386*, 116–124. [CrossRef]
4. Ghasemlou, M.; Daver, F.; Ivanova, E.P.; Adhikari, B. Bio-inspired sustainable and durable superhydrophobic materials: From nature to market. *J. Mater. Chem. A* **2019**, *7*, 16643–16670. [CrossRef]
5. Young, T., III. An essay on the cohesion of fluids. *Philos. Trans. R. Soc. Lond.* **1805**, *95*, 65–87.
6. Wenzel, R.N. Resistance of solid surfaces to wetting by water. *Ind. Eng. Chem.* **1936**, *28*, 988–994. [CrossRef]
7. Cassie, A.B.D.; Baxter, S. Wettability of Porous Surfaces. *Trans. Faraday Soc.* **1944**, *40*, 546–551. [CrossRef]

8. Huang, J.-Y.; Wang, F.-H.; Zhao, X.; Zhang, K. Wetting Transition and Stability Testing of Superhydrophobic State. *Acta Phys. -Chim. Sin.* **2013**, *29*, 2459–2464.
9. Lafuma, A.; Quéré, D. Superhydrophobic states. *Nat. Mater.* **2003**, *2*, 457–460. [CrossRef]
10. Bormashenko, E.; Pogreb, R.; Whyman, G.; Erlich, M. Resonance Cassie−Wenzel Wetting Transition for Horizontally Vibrated Drops Deposited on a Rough Surface. *Langmuir* **2007**, *23*, 12217–12221. [CrossRef]
11. Bartolo, D.; Bouamrirene, F.; Verneuil, E.; Buguin, A.; Silberzan, P.; Moulinet, S. Bouncing or sticky droplets: Impalement transitions on superhydrophobic micropatterned surfaces. *Eur. Lett.* **2006**, *74*, 299–305. [CrossRef]
12. Kwon, H.-M.; Paxson, A.T.; Varanasi, K.K.; Patankar, N.A. Rapid Deceleration-Driven Wetting Transition during Pendant Drop Deposition on Superhydrophobic Surfaces. *Phys. Rev. Lett.* **2011**, *106*, 036102. [CrossRef]
13. Naylor, R.S.D. *Cause Experiment and Science: A Galilean Dialogue Incorporating a New English Translation of Galileo's 'Bodies That Stay Atop Water or Move in it'*; University of Chicago Press: Chicago, IL, USA, 1981.
14. Kota, A.K.; Mabry, J.M.; Tuteja, A. Superoleophobic surfaces: Design criteria and recent studies. *Surf. Innov.* **2013**, *1*, 71–83. [CrossRef]
15. Xue, Y.; Chu, S.; Lv, P.; Duan, H. Importance of Hierarchical Structures in Wetting Stability on Submersed Superhydrophobic Surfaces. *Langmuir* **2012**, *28*, 9440–9450. [CrossRef] [PubMed]
16. Poetes, R.; Holtzmann, K.; Franze, K.; Steiner, U. Metastable Underwater Superhydrophobicity. *Phys. Rev. Lett.* **2010**, *105*, 166104. [CrossRef] [PubMed]
17. Lv, P.; Xue, Y.; Shi, Y.; Lin, H.; Duan, H. Metastable states and wetting transition of submerged superhydrophobic structures. *Phys. Rev. Lett.* **2014**, *112*, 196101. [CrossRef] [PubMed]
18. Zhao, M.; Zhang, Z.; Ahn, Y.; Jang, J. Molecular simulation study on the wettability of a surface corrugated with trapezoidal nanopillars. *Appl. Surf. Sci.* **2020**, *537*, 147918. [CrossRef]
19. Zhang, Z.; Zhao, M.; Ahn, Y.; Jang, J. Wettability of a surface engraved with the periodic nanoscale trenches: Effects of geometry and pressure. *J. Mol. Liq.* **2021**, *335*, 116276. [CrossRef]
20. Bai, L.; Kim, K.; Ha, M.Y.; Ahn, Y.; Jang, J. Molecular Insights on the Wetting Behavior of a Surface Corrugated with Nanoscale Domed Pillars. *Langmuir* **2021**, *37*, 9336–9345. [CrossRef]
21. Yan, M.; Li, T.; Zheng, P.; Wei, R.; Jiang, Y.; Li, H. Wetting state transition of a liquid gallium drop at the nanoscale. *Phys. Chem. Chem. Phys.* **2020**, *22*, 11809–11816. [CrossRef]
22. Zhang, M.; Ma, L.; Wang, Q.; Hao, P.; Zheng, X. Wettability behavior of nanodroplets on copper surfaces with hierarchical nanostructures. *Colloids Surf. A Physicochem. Eng. Asp.* **2020**, *604*, 125291. [CrossRef]
23. Li, H.; Zhang, K. Dynamic behavior of water droplets impacting on the superhydrophobic surface: Both experimental study and molecular dynamics simulation study. *Appl. Surf. Sci.* **2019**, *498*, 143793. [CrossRef]
24. Han, X.; Wang, M.; Yan, R.; Wang, H. Cassie State Stability and Gas Restoration Capability of Superhydrophobic Surfaces with Truncated Cone-Shaped Pillars. *Langmuir* **2021**, *37*, 12897–12906. [CrossRef] [PubMed]
25. Hosseini, S.; Savaloni, H.; Shahraki, M.G. Influence of surface morphology and nano-structure on hydrophobicity: A molecular dynamics approach. *Appl. Surf. Sci.* **2019**, *485*, 536–546. [CrossRef]
26. Saleki, O.; Moosavi, A.; Hannani, S.K. Molecular Dynamics Study of Friction Reduction of Two-Phase Flows on Surfaces Using 3D Hierarchical Nanostructures. *J. Phys. Chem. C* **2019**, *123*, 27519–27530. [CrossRef]
27. Li, H.; Yan, T. Importance of moderate size of pillars and dual-scale structures for stable superhydrophobic surfaces: A molecular dynamics simulation study. *Comput. Mater. Sci.* **2020**, *175*, 109613. [CrossRef]
28. Zhang, B.-X.; Wang, S.-L.; Wang, Y.-B.; Yang, Y.-R.; Wang, X.-D.; Yang, R.-G. Harnessing Reversible Wetting Transition to Sweep Contaminated Superhydrophobic Surfaces. *Langmuir* **2021**, *37*, 3929–3938. [CrossRef]
29. Bryk, P.; Terzyk, A.P. Chasing the Critical Wetting Transition. An Effective Interface Potential Method. *Materials* **2021**, *14*, 7138. [CrossRef]
30. Lopes, D.M.; Mombach, J.C.M. Two-Dimensional Wetting Transition Modeling with the Potts Model. *Braz. J. Phys.* **2017**, *47*, 672–677. [CrossRef]
31. Yin, B.; Xu, S.; Yang, S.; Dong, F. Shape Optimization of a Microhole Surface for Control of Droplet Wettability via the Lattice Boltzmann Method and Response Surface Methodology. *Langmuir* **2021**, *37*, 3620–3627. [CrossRef]
32. Zhou, W.; Yan, Y.; Liu, X.; Liu, B. Lattice Boltzmann parallel simulation of microflow dynamics over structured surfaces. *Adv. Eng. Softw.* **2017**, *107*, 51–58. [CrossRef]
33. Gong, W.; Zu, Y.; Chen, S.; Yan, Y. Wetting transition energy curves for a droplet on a square-post patterned surface. *Sci. Bull.* **2016**, *62*, 136–142. [CrossRef]
34. Gong, W.; Yan, Y.; Chen, S.; Giddings, D. Numerical study of wetting transitions on biomimetic surfaces using a lattice Boltzmann approach with large density ratio. *J. Bionic Eng.* **2017**, *14*, 486–496. [CrossRef]
35. Farokhirad, S.; Lee, T. Computational study of microparticle effect on self-propelled jumping of droplets from superhydrophobic substrates. *Int. J. Multiph. Flow* **2017**, *95*, 220–234. [CrossRef]
36. Fakhari, A.; Bolster, D. Diffuse interface modeling of three-phase contact line dynamics on curved boundaries: A lattice Boltzmann model for large density and viscosity ratios. *J. Comput. Phys.* **2017**, *334*, 620–638. [CrossRef]
37. Dou, S.; Hao, L. Investigation of wetting states and wetting transition of droplets on the microstructured surface using the lattice Boltzmann model. *Numer. Heat Transfer. Part A Appl.* **2020**, *78*, 321–337. [CrossRef]

38. Randive, P.; Dalal, A.; Mukherjee, P.P. Probing the influence of confinement and wettability on droplet displacement behavior: A mesoscale analysis. *Eur. J. Mech. B Fluids* **2018**, *75*, 327–338. [CrossRef]
39. Sumit, P.; Prakhar, D.; Sujay, C. Superhydrophobic Surfaces: Insignts from Theory and Experiment. *J. Phys. Chem.* **2020**, *124*, 1323–1360.
40. McHale, G. Cassie and Wenzel: Were They Really So Wrong? *Langmuir* **2007**, *23*, 8200–8205. [CrossRef]
41. Greenwood, J.A.; Williamson, J.B.P. Contact of nominally flat surfaces. *Proc. R. Soc. Lond. Ser. A Math. Phys. Sci.* **1966**, *295*, 300–319.
42. Jiang, Y.; Choi, C. Droplet rentention on superhydrophobic surfaces: A Critical Review. *Adv. Mater. Interfaces* **2020**, *8*, 2001205. [CrossRef]
43. Kim, S.; Kim, D.H.; Su, H.C.; Woo, Y.K.; Sin, K.; Young, T.C. Effect of surface pattern morphology on inducing superhydrophobicity. *Appl. Surf. Sci.* **2020**, *513*, 145847. [CrossRef]
44. Cao, J.; Gao, D.; Li, C.; Si, X.; Jia, J.; Qi, J. Bioinspired Metal-Intermetallic Laminated Composites for the Frabrication of Superhydrophobic Surfaces with Responsive Wettability. *Acs Appl. Mater. Interfaces* **2021**, *13*, 5834–5843. [CrossRef] [PubMed]
45. Mandelbrot, B. How Long is the Coast of Britain? Statistical Self-Similarity and Fractional Dimension. *Science* **1967**, *156*, 636–638. [CrossRef] [PubMed]
46. Love, J.C.; Gates, B.D.; Wolfe, D.B.; Paul, K.E.; Whitesides, G.M. Fabrication and wetting properties of metallic half-shells with submicron diameters. *Nano Lett.* **2002**, *2*, 891. [CrossRef]
47. Qian, X.-S. Modern Mechanics-A Speech at the 1978 National Conference on Mechanical Planning. *Mech. Eng.* **1979**, 4–9.
48. Shibuichi, S.; Onda, T.; Satoh, N.; Tsujii, K. Super Water-Repellent Surfaces Resulting from Fractal Structure. *J. Phys. Chem.* **1996**, *100*, 19512–19517. [CrossRef]
49. Synytska, A.; Ionov, L.; Grundke, K.; Stamm, M. Wetting on Fractal Superhydrophobic Surfaces from "Core-shell" Particles: A Comparison of Theory and Experiment. *Langmuir* **2009**, *25*, 3132–3136. [CrossRef]
50. Lin, Y.; Zhou, R.; Xu, J. Superhydrophobic Surfaces Based on Fractal and Hierarchical Microstructures Using Two-Photon Polymerization: Toward Flexible Superhydrophobic Films. *Adv. Mater. Interfaces* **2018**, *5*, 1801126. [CrossRef]
51. Davis, E.; Liu, Y.; Jiang, L.; Lu, Y.; Ndao, S. Wetting characteristics of 3-dimensional nanostructured fractal surfaces. *Appl. Surf. Sci.* **2017**, *392*, 929–935. [CrossRef]
52. Tuteja, A.; Choi, W.; Mabry, J.M.; McKinley, G.H.; Cohen, R.E. COLL 19-Designing superoleophobic surfaces with fluoroPOSS. *Abstr. Pap. Am. Chem. Soc.* **2007**, *234*.
53. Ahuja, A.; Taylor, J.A.; Lifton, V.; Sidorenko, A.A.; Salamon, T.R.; Lobaton, E.J.; Kolodner, P.; Krupenkin, T.N. Nanonails: A simple geometrical approach to electrically tunable superlyophobic surfaces. *Langmuir* **2008**, *24*, 9–14. [CrossRef] [PubMed]
54. Li, Y.; Han, X.; Jin, H.; Li, W. Understanding superhydrophobic behaviors on hydrophilic materials: A thermodynamic approach. *Mater. Res. Express* **2021**, *8*, 076403. [CrossRef]
55. Pang, X.; Hao, L. Research on the Wettability of Hemisphere Surface. *Mater. Rev. B* **2013**, *27*, 150–153.
56. Kwon, Y.; Choi, S.; Anantharaju, N.; Lee, J.; Panchagnula, M.V.; Patankar, N.A. Is the Cassie–Baxter Formula Relevant? *Langmuir* **2010**, *26*, 17528–17531. [CrossRef] [PubMed]
57. Gao, L.; McCarthy, T.J. How Wenzel and Cassie Were Wrong. *Langmuir* **2007**, *23*, 3762–3765. [CrossRef]
58. Extrand, C.W. Model for Contact Angles and Hysteresis on Rough and Ultraphobic Surfaces. *Langmuir* **2002**, *18*, 7991–7999. [CrossRef]
59. Öner, D.; McCarthy, T.J. Ultrahydrophobic Surfaces. Effects of Topography Length Scales on Wettability. *Langmuir* **2000**, *16*, 7777–7782. [CrossRef]
60. Butt, H.-J.; Roisman, I.; Brinkmann, M.; Papadopoulos, P.; Vollmer, D.; Semprebon, C. Characterization of super liquid-repellent surfaces. *Curr. Opin. Colloid Interface Sci.* **2014**, *19*, 343–354. [CrossRef]
61. Papadopoulos, P.; Mammen, L.; Deng, X.; Vollmer, D.; Butt, H.J. How superhydrophobicity breaks down. In Proceedings of the National Academy of Sciences of the United States of America, Princeton, NJ, USA, 4 February 2013; pp. 3254–3258.
62. Teisala, H.; Butt, H.-J. Hierarchical Structures for Superhydrophobic and Superoleophobic Surfaces. *Langmuir* **2018**, *35*, 10689–10703. [CrossRef]
63. Teisala, H.; Geyer, F.; Haapanen, J.; Juuti, P.; Mäkelä, J.M.; Vollmer, D.; Butt, H.-J. Ultrafast Processing of Hierarchical Nanotexture for a Transparent Superamphiphobic Coating with Extremely Low Roll-Off Angle and High Impalement Pressure. *Adv. Mater.* **2018**, *30*, e1706529. [CrossRef]
64. Tretyakov, N.; Papadopoulos, P.; Vollmer, D.; Duenweg, B.; Daoulas, K.C.; Butt, H.-J. The Cassie-Wenzel transition of fluids on nanostructured substrates: Macroscopic force balance versus microscopic density-functional theory. *J. Chem. Phys.* **2016**, *145*, 134703. [CrossRef] [PubMed]
65. Patankar, N.A. Transition between Superhydrophobic States on Rough Surfaces. *Langmuir* **2004**, *20*, 7097–7102. [CrossRef] [PubMed]
66. Patankar, N.A. Consolidation of Hydrophobic Transition Criteria by Using an Approximate Energy Minimization Approach. *Langmuir* **2010**, *26*, 8941–8945. [CrossRef] [PubMed]
67. Deng, T.; Varanasi, K.K.; Hsu, M.; Bhate, N.; Keimel, C.; Stein, J.; Blohm, M.L. Nonwetting of impinging droplets on textured surfaces. *Appl. Phys. Lett.* **2009**, *94*, 133109. [CrossRef]

68. Extrand, C.W. Contact Angles and Their Hysteresis as a Measure of Liquid−Solid Adhesion. *Langmuir* **2004**, *20*, 4017–4021. [CrossRef] [PubMed]
69. Jung, Y.C.; Bhushan, B. Wetting transition of water droplets on superhydrophobic patterned surfaces. *Scr. Mater.* **2007**, *57*, 1057–1060. [CrossRef]
70. Kusumaatmaja, H.; Blow, M.; Dupuis, A.; Yeomans, J. The collapse transition on superhydrophobic surfaces. *Eur. Lett.* **2008**, *81*, 36003. [CrossRef]
71. Reyssat, M.; Pépin, A.; Marty, F.; Chen, Y.; Quéré, D. Bouncing transitions on microtextured materials. *Eur. Lett.* **2006**, *74*, 306–312. [CrossRef]
72. Ren, W. Wetting Transition on Patterned Surfaces: Transition States and Energy Barriers. *Langmuir* **2014**, *30*, 2879–2885. [CrossRef]
73. Whyman, G.; Bormashenko, E. How to Make the Cassie Wetting State Stable? *Langmuir* **2011**, *27*, 8171–8176. [CrossRef]
74. Dupre, A. *Theorie Mecanique de La Chaleur*; Gauthier-Villars: Pairs, France, 1869; pp. 8941–8945.
75. Zheng, Q.-S.; Yu, Y.; Zhao, Z.-H. Effects of Hydraulic Pressure on the Stability and Transition of Wetting Modes of Superhydrophobic Surfaces. *Langmuir* **2005**, *21*, 12207–12212. [CrossRef] [PubMed]
76. Extrand, C.W. Modeling of Ultralyophobicity: Suspension of Liquid Drops by a Single Asperity. *Langmuir* **2005**, *21*, 10370–10374. [CrossRef] [PubMed]
77. Extrand, C.W. Designing for Optimum Liquid Repellency. *Langmuir* **2006**, *22*, 1711–1714. [CrossRef] [PubMed]
78. Amirfazli, A.; Neumann, A.W. Status of the three-phase line tension. *Adv. Colloid Interface Sci.* **2004**, *110*, 121–141. [CrossRef] [PubMed]
79. Marmur, A. Line Tension and the Intrinsic Contact Angle in Solid–Liquid–Fluid Systems. *J. Colloid Interface Sci.* **1997**, *186*, 462–466. [CrossRef]
80. Checco, A.; Guenoun, P.; Daillant, J. Nonlinear Dependence of the Contact Angle of Nanodroplets on Contact Line Curvature. *Phys. Rev. Lett.* **2003**, *91*, 186101. [CrossRef]
81. Pompe, T.; Fery, A.; Herminghaus, S. Measurement of contact line tension by analysis of the three-phase boundary with nanometer resolution. In Proceedings of the International Symposium on Apparent and Microscopic Contact Angles in Conjunction with the 216th American-Chemical-Society Meeting, Boston, MA, USA, 23 August 1998; pp. 3–12.
82. Lei, L.; Li, H.; Shi, J.; Chen, Y. Diffraction Patterns of a Water-Submerged Superhydrophobic Grating under Pressure. *Langmuir* **2009**, *26*, 3666–3669. [CrossRef]
83. Priest, C.; Albrecht, T.W.J.; Sedev, R.; Ralston, J. Asymmetric Wetting Hysteresis on Hydrophobic Microstructured Surfaces. *Langmuir* **2009**, *25*, 5655–5660. [CrossRef]
84. Anantharaju, N.; Panchagnula, M.V.; Vedantam, S. Asymmetric Wetting of Patterned Surfaces Composed of Intrinsically Hysteretic Materials. *Langmuir* **2009**, *25*, 7410–7415. [CrossRef]
85. Lv, P.; Xue, Y.; Liu, H.; Shi, Y.; Xi, P.; Lin, H.; Duan, H. Symmetric and Asymmetric Meniscus Collapse in Wetting Transition on Submerged Structured Surfaces. *Langmuir* **2014**, *31*, 1248–1254. [CrossRef]
86. Fetzer, R.; Ralston, J. Exploring Defect Height and Angle on Asymmetric Contact Line Pinning. *J. Phys. Chem. C* **2011**, *115*, 14907–14913. [CrossRef]
87. Giacomello, A.; Chinappi, M.; Meloni, S.; Casciola, C.M. Metastable Wetting on Superhydrophobic Surfaces: Continuum and Atomistic Views of the Cassie-Baxter–Wenzel Transition. *Phys. Rev. Lett.* **2012**, *109*, 226102. [CrossRef] [PubMed]
88. Giacomello, A.; Meloni, S.; Chinappi, M.; Casciola, C.M. Cassie-Baxter and Wenzel States on a Nanostructured Surface: Phase Diagram, Metastabilities, and Transition Mechanism by Atomistic Free Energy Calculations. *Langmuir* **2012**, *28*, 10764–10772. [CrossRef] [PubMed]
89. Kim, H.; Ha, M.Y.; Jang, J. Wetting Transition of a Cylindrical Cavity Engraved on a Hydrophobic Surface. *J. Phys. Chem. C* **2018**, *122*, 2122–2126. [CrossRef]
90. Byun, D.; Hong, J.; Saputra; Ko, J.H.; Lee, Y.J.; Park, H.C.; Byun, B.-K.; Lukes, J.R. Wetting Characteristics of Insect Wing Surfaces. *J. Bionic Eng.* **2009**, *6*, 63–70. [CrossRef]
91. Bormashenko, E.; Gendelman, O.; Whyman, G. Superhydrophobicity of Lotus Leaves versus Birds Wings: Different Physical Mechanisms Leading to Similar Phenomena. *Langmuir* **2012**, *28*, 14992–14997. [CrossRef]
92. Zu, Y.Q.; Yan, Y.Y.; Li, J.Q.; Han, Z.W. Wetting Behaviours of a Single Droplet on Biomimetic Micro Structured Surfaces. *J. Bionic Eng.* **2010**, *7*, 191–198. [CrossRef]
93. Yin, L.; Zhu, L.; Wang, Q.; Ding, J.; Chen, Q. Superhydrophobicity of Natural and Artificial Surfaces under Controlled Condensation Conditions. *ACS Appl. Mater. Interfaces* **2011**, *3*, 1254–1260. [CrossRef]
94. Ju, L.; Xiao, H.; Ye, L.; Hu, A.; Li, M. Wettability evolution of different nanostructured cobalt films based on electrodeposition. *Micro Nano Lett.* **2017**, *12*, 470–473. [CrossRef]
95. Vüllers, F.; Peppou-Chapman, S.; Kavalenka, M.N.; Hölscher, H.; Neto, C. Effect of repeated immersions and contamination on plastron stability in superhydrophobic surfaces. *Phys. Fluids* **2019**, *31*, 012102. [CrossRef]
96. Bormashenko, E. Progress in understanding wetting transitions on rough surfaces. *Adv. Colloid Interface Sci.* **2015**, *222*, 92–103. [CrossRef] [PubMed]
97. Hemeda, A.A.; Gad-El-Hak, M.; Tafreshi, H.V. Effects of hierarchical features on longevity of submerged superhydrophobic surfaces with parallel grooves. *Phys. Fluids* **2014**, *26*, 082103. [CrossRef]

98. Zhang, B.; Wang, J.; Zhang, X. Effects of the Hierarchical Structure of Rough Solid Surfaces on the Wetting of Microdroplets. *Langmuir* **2013**, *29*, 6652–6658. [CrossRef]
99. Lee, J.S.; Lee, J.S. Multiphase static droplet simulations in hierarchically structured super-hydrophobic surfaces. *J. Mech. Sci. Technol.* **2016**, *30*, 3741–3747. [CrossRef]
100. Shen, Y.; Tao, J.; Tao, H.; Chen, S.; Pan, L.; Wang, T. Nanostructures in superhydrophobic Ti6Al4V hierarchical surfaces control wetting state transitions. *Soft Matter* **2015**, *11*, 3806–3811. [CrossRef]
101. Teisala, H.; Tuominen, M.; Aromaa, M.; Stepien, M.; Mäkelä, J.M.; Saarinen, J.J.; Toivakka, M.; Kuusipalo, J. Nanostructures Increase Water Droplet Adhesion on Hierarchically Rough Superhydrophobic Surfaces. *Langmuir* **2012**, *28*, 3138–3145. [CrossRef]
102. Liang, W.; He, L.; Wang, F.; Yang, B.; Wang, Z. A 3-D model for thermodynamic analysis of hierarchical structured superhydrophobic surfaces. *Colloids Surf. A Physicochem. Eng. Asp.* **2017**, *523*, 98–105. [CrossRef]
103. Nosonovsky, M.; Bhushan, B. Superhydrophobic surfaces and emerging applications: Non-adhesion, energy, green engineering. *Curr. Opin. Colloid Interface Sci.* **2009**, *14*, 270–280. [CrossRef]
104. Hejazi, V.; Nosonovsky, M. Contact angle hysteresis in multiphase systems. *Colloid Polym. Sci.* **2012**, *291*, 329–338. [CrossRef]
105. Cai, M.; Li, Y.; Chen, Y.; Xu, J.; Zhang, L.; Lei, J. Wettability Transition of a Liquid Droplet on Solid Surface With Nanoscale Inverted Triangular Grooves. In Proceedings of the ASME 2019 6th International Conference on Micro/Nanoscale Heat and Mass Transfer, Dalian, China, 8–10 July 2019.
106. Dufour, R.; Saad, N.; Carlier, J.; Campistron, P.; Nassar, G.; Toubal, M.; Boukherroub, R.; Senez, V.; Nongaillard, B.; Thomy, V. Acoustic Tracking of Cassie to Wenzel Wetting Transitions. *Langmuir* **2013**, *29*, 13129–13134. [CrossRef]
107. Liu, Y.; Xiu, Y.; Hess, D.W.; Wong, C.P. Silicon surface structure-controlled olephobicity. *Langmuir* **2010**, *26*, 8908–8913. [CrossRef] [PubMed]
108. Savoy, E.S.; Escobedo, F.A. Simulation Study of Free-Energy Barriers in the Wetting Transition of an Oily Fluid on a Rough Surface with Reentrant Geometry. *Langmuir* **2012**, *28*, 16080–16090. [CrossRef] [PubMed]
109. Wang, Z.; Lin, K.; Zhao, Y.-P. The effect of sharp solid edges on the droplet wettability. *J. Colloid Interface Sci.* **2019**, *552*, 563–571. [CrossRef]
110. Bobji, M.S.; Kumar, S.V.; Asthana, A.N.R. Govardhan, Underwater Sustainability of the "Cassie" State of Wetting. *Langmuir* **2009**, *25*, 12120–12126. [CrossRef] [PubMed]
111. Forsberg, P.; Nikolajeff, F.; Karlsson, M. Cassie–Wenzel and Wenzel–Cassie transitions on immersed superhydrophobic surfaces under hydrostatic pressure. *Soft Matter* **2010**, *7*, 104–109. [CrossRef]
112. Moulinet, S.; Bartolo, D. Life and death of a fakir droplet: Impalement transitions on superhydrophobic surfaces. *Eur. Phys. J. E* **2007**, *24*, 251–260. [CrossRef]
113. Sakai, M.; Yanagisawa, T.; Nakajima, A.; Kameshima, Y.; Okada, K. Effect of Surface Structure on the Sustainability of an Air Layer on Superhydrophobic Coatings in a Water−Ethanol Mixture. *Langmuir* **2008**, *25*, 13–16. [CrossRef]
114. Xue, Y.; Lv, P.; Lin, H.; Duan, H. Underwater Superhydrophobicity: Stability, Design and Regulation, and Applications. *Appl. Mech. Rev.* **2016**, *68*, 030803. [CrossRef]
115. Rathgen, H.; Mugele, F. Microscopic shape and contact angle measurement at a superhydrophobic surface. *Faraday Discuss.* **2010**, *146*, 49–56. [CrossRef]
116. Luo, C.; Zheng, H.; Wang, L.; Fang, H.; Hu, J.; Fan, C.; Cao, Y.; Wang, J. Direct Three-Dimensional Imaging of the Buried Interfaces between Water and Superhydrophobic Surfaces. *Angew. Chem. Int. Ed.* **2010**, *49*, 9145–9148. [CrossRef]
117. Papadopoulos, P.; Deng, X.; Mammen, L.; Drotlef, D.-M.; Battagliarin, G.; Li, C.; Müllen, K.; Landfester, K.; Del Campo, A.; Butt, H.-J.; et al. Wetting on the Microscale: Shape of a Liquid Drop on a Microstructured Surface at Different Length Scales. *Langmuir* **2012**, *28*, 8392–8398. [CrossRef] [PubMed]
118. Tsai, P.; Pacheco, S.; Pirat, C.; Lefferts, L.; Lohse, D. Drop Impact upon Micro- and Nanostructured Superhydrophobic Surfaces. *Langmuir* **2009**, *25*, 12293–12298. [CrossRef] [PubMed]
119. Deng, X.; Schellenberger, F.; Papadopoulos, P.; Vollmer, D.; Butt, H.-J. Liquid Drops Impacting Superamphiphobic Coatings. *Langmuir* **2013**, *29*, 7847–7856. [CrossRef] [PubMed]
120. Hao, C.; Li, J.; Liu, Y.; Zhou, X.; Liu, Y.; Liu, R.; Che, L.; Zhou, W.; Sun, D.; Li, L.; et al. Superhydrophobic-like tunable droplet bouncing on slippery liquid interfaces. *Nat. Commun.* **2015**, *6*, 7986. [CrossRef] [PubMed]
121. Cannon, A.H.; King, W.P. Visualizing contact line phenomena on microstructured superhydrophobic surfaces. *J. Vac. Sci. Technol. B* **2010**, *28*, L21–L24. [CrossRef]
122. Rykaczewski, K.; Landin, T.; Walker, M.L.; Scott, J.H.J.; Varanasi, K.K. Direct Imaging of Complex Nano- to Microscale Interfaces Involving Solid, Liquid, and Gas Phases. *ACS Nano* **2012**, *6*, 9326–9334. [CrossRef]
123. Wiedemann, S.; Plettl, A.; Walther, P.; Ziemann, P. Freeze Fracture Approach to Directly Visualize Wetting Transitions on Nanopatterned Superhydrophobic Silicon Surfaces: More than a Proof of Principle. *Langmuir* **2013**, *29*, 913–919. [CrossRef]
124. Li, S.; Lamant, S.; Carlier, J.; Toubal, M.; Campistron, P.; Xu, X.; Vereecke, G.; Senez, V.; Thomy, V.; Nongaillard, B. High-Frequency Acoustic for Nanostructure Wetting Characterization. *Langmuir* **2014**, *30*, 7601–7608. [CrossRef]
125. Su, J.; Esmaeilzadeh, H.; Wang, P.; Ji, S.; Inalpolat, M.; Charmchi, M.; Sun, H. Effect of wetting states on frequency response of a micropillar-based quartz crystal microbalance. *Sens. Actuators A Phys.* **2018**, *286*, 115–122. [CrossRef]

126. Sudeepthi, A.; Yeo, L.; Sen, A.K. Cassie–Wenzel wetting transition on nanostructured superhydrophobic surfaces induced by surface acoustic waves. *Appl. Phys. Lett.* **2020**, *116*, 093704. [CrossRef]
127. Ensikat, H.J.; Schulte, A.J.; Koch, K.; Barthlott, W. Droplets on Superhydrophobic Surfaces: Visualization of the Contact Area by Cryo-Scanning Electron Microscopy. *Langmuir* **2009**, *25*, 13077–13083. [CrossRef] [PubMed]
128. Li, X.; Ma, X.; Lan, Z. Dynamic Behavior of the Water Droplet Impact on a Textured Hydrophobic/Superhydrophobic Surface: The Effect of the Remaining Liquid Film Arising on the Pillars' Tops on the Contact Time. *Langmuir* **2010**, *26*, 4831–4838. [CrossRef] [PubMed]

Article

Virucidal and Bactericidal Filtration Media from Electrospun Polylactic Acid Nanofibres Capable of Protecting against COVID-19

Fabrice Noël Hakan Karabulut *, Dhevesh Fomra *, Günther Höfler, Naveen Ashok Chand and Gareth Wesley Beckermann

NanoLayr Ltd., 59 Mahunga Drive, Mangere Bridge, Auckland 2022, New Zealand; gunther.hofler@nanolayr.com (G.H.); naveen.chand@nanolayr.com (N.A.C.); gareth.beckermann@nanolayr.com (G.W.B.)
* Correspondence: fabrice.karabulut@nanolayr.com (F.N.H.K.); dhevesh.fomra@nanolayr.com (D.F.)

Abstract: Electrospun nanofibres excel at air filtration owing to diverse filtration mechanisms, thereby outperforming meltblown fibres. In this work, we present an electrospun polylactide acid nanofibre filter media, FilterLayr™ Eco, displaying outstanding bactericidal and virucidal properties using Manuka oil. Given the existing COVID-19 pandemic, face masks are now a mandatory accessory in many countries, and at the same time, they have become a source of environmental pollution. Made by NanoLayr Ltd., FilterLayr™ Eco uses biobased renewable raw materials with products that have end-of-life options for being industrially compostable. Loaded with natural and non-toxic terpenoid from manuka oil, FilterLayr Eco can filter up to 99.9% of 0.1 μm particles and kill >99% of trapped airborne fungi, bacteria, and viruses, including SARS-CoV-2 (Delta variant). In addition, the antimicrobial activity, and the efficacy of the filter media to filtrate particles was shown to remain highly active following several washing cycles, making it a reusable and more environmentally friendly option. The new nanofibre filter media, FilterLayr™ Eco, met the particle filtration efficiency and breathability requirements of the following standards: N95 performance in accordance with NIOSH 42CFR84, level 2 performance in accordance with ASTM F2100, and level 2 filtration efficiency and level 1 breathability in accordance with ASTM F3502. These are globally recognized facemask and respirator standards.

Keywords: electrospun; nanofibre; polylactid acid; filtration; SARS-CoV-2; filter media

Citation: Karabulut, F.N.H.; Fomra, D.; Höfler, G.; Chand, N.A.; Beckermann, G.W. Virucidal and Bactericidal Filtration Media from Electrospun Polylactic Acid Nanofibres Capable of Protecting against COVID-19. *Membranes* **2022**, *12*, 571. https://doi.org/10.3390/membranes12060571

Academic Editors: Maria Vittoria Diamanti, Massimiliano D'Arienzo, Carlo Antonini and Michele Ferrari

Received: 21 April 2022
Accepted: 24 May 2022
Published: 30 May 2022

Publisher's Note: MDPI stays neutral with regard to jurisdictional claims in published maps and institutional affiliations.

Copyright: © 2022 by the authors. Licensee MDPI, Basel, Switzerland. This article is an open access article distributed under the terms and conditions of the Creative Commons Attribution (CC BY) license (https://creativecommons.org/licenses/by/4.0/).

1. Introduction

The recent worldwide pandemic, caused by the respiratory disease SARS-CoV-2, has shed light on the importance of air quality and excellent air filtration. Awareness of the impact that air quality has on human health has led to an increase in the demand for better personal protection against airborne particles and microorganisms. Personal protective equipment (PPE) is the most accepted method of self-protection in situations where safe social distancing is not possible. Masks and respirators are undoubtedly the most critical pieces of protective equipment for preventing the spread of particulate matter such as fungi, bacteria, and viruses [1–4]. However, many concerns remain about the survival of microbes on the face mask surface [5–7] re-aerosolisation of settled particles [8], proper management and disposal of old face masks [9], and fomite transmission [10]. In order to impart surface contact killing/deactivation mechanisms in addition to particle filtration, some of these concerns have led to the development of face masks with inherent antimicrobial capabilities. Most face masks contain plastics or similar materials, making their widespread and ever-increasing use a significant environmental issue. In a short span of time, mask usage during COVID-19 has generated millions of tons of plastic waste.

When utilised as a protective filter media layer in facemasks, nanofibres (NFs) have unique properties [11–13]. The electrospinning (ES) method is considered to be the best for generating polymeric NFs, as a wide range of synthetic and or bio-based polymers can be used at both the lab and commercial scale [14]. Last year, Beckermann et al. introduced the ability of electrospun PMMA/EVOH NF filter media to pass stringent international standards [11]. However, the authors reported a filter media made of a polymer blend (PMMA/EVOH); materials which are not considered sustainable. Due to the COVID-19 pandemic, the monthly usage of single-use facemasks during 2021 was estimated at 129 billion [15], Ref. [15] resulting in the generation of a lot of waste, much of which inevitably ended up in the environment. Recent trends drive research and development to focus on biodegradability, compostability and reusability of the next generation of face masks.

ES technique can be used to produce functional NFs with enhanced antimicrobial properties when loaded with antimicrobial agents [16–18]. Additives such as silver (Ag) [19–22] and copper (Cu) nanoparticles (NPs) [23] and plant extracts such as *Centella asiatica*, *Myristica andamanica*, aloe vera and grape seed [24–28] have demonstrated effective antimicrobial properties against a wide range of pathogens and can be incorporated into the electrospun formulations. Currently, many antimicrobial textiles are coated with NPs of metals such as copper and silver [29–36]. These NPs provide antimicrobial properties by releasing Ag^+ or Cu^{2+} metallic ions or reactive oxygen species, triggering bacterial cell-wall damage [37]. These metal species, if loosened from the materials, could leak into grey water (after washing) or onto the consumer's skin through wearing. Antimicrobial face masks containing copper, silver, or other antimicrobials are widely advertised. The methods for loading antimicrobials onto facemask fibres are often undisclosed, and the risk of metal leaching from these masks is unknown. Recently, Goldfard et al. reported the quantity of silver and copper metals leached from nine commercially available 'antimicrobial' masks and a 100% cotton face mask used as a control. These masks were exposed to laundry detergent, artificial saliva, and deionized water to mimic normal facemask usage and care. The authors observed that leaching differed significantly depending on the brand, leaching solution and metal NPs. After an hour of exposure, in some cases, the full content of the metals contained in the original face mask leached [38]. Findings such as these have prompted researchers to investigate naturally derived bioactive agents as more favourable alternatives to heavy metals.

To address both the viral threats and environmental impacts, the development of a unique face mask filter medium is described in this paper, made from electrospun bio-based polylactic acid (PLA). This filter layer boasts an additional antimicrobial functionality imparted by the addition of non-toxic and naturally sourced terpenoids extracted from manuka oil. The antimicrobial properties of manuka triketone have been previously reported in applications such as wound dressing [39] but have not been demonstrated as an active component of face masks. This novel PLA filter media successfully achieved high filtration efficiency for particulate pollutants and pathogens. The PLA ES NF filter media were tested according to the following international standards: ASTM Test Method F2299 [40], ASTM Test Method F3502 [41] and NIOSH 42CFR84 (N95) [42]. Additionally, the antimicrobial performance of this filter media were evaluated against different bacteria strains (*Escherichia coli*, *Staphylococcus aureus*, and *Klebsiella pneumoniae* following the test standard ISO 20743:2013 [43]), viruses (influenza A (H1N1), human coronavirus and SARS-CoV-2 following the test standard ISO 18184:2019 [44]) and a fungus (*Aspergillus niger*) following the test standard ASTM G21 [45]). Lastly, to test the reusability of the filter media after being subjected to 5–10 laundering cycles, the particulate filtration efficiency and antimicrobial activity were tested again according to ASTM F3502 [41] and ISO 20743:2013 [43] standards.

2. Experimental

2.1. Materials

The following materials have been used to prepare solutions for electrospinning: polylactic acid (PLA); formic acid and acetic acid purchased from Merck (Kenilworth, NJ, USA). Manuka oil was purchased from a local New Zealand supplier. To protect the nanofibre layer, spun-bonded PLA, a non-woven fabric, was used as a substrate and cover layer.

2.2. Electrospinning of Nanofibre and Characterisation

The most widely used method to produce polymer NFs is electrospinning. This has been explained in detail by Rutledge and Fridrikh [46]. Nanolayr Ltd. (Auckland, New Zealand) formerly known as Revolution Fibres Ltd., has developed and patented a revolutionary needleless electrospinning process which was used to manufacture this filter media in roll form, and which was used in this investigation. The solutions used in the ES process were made by dissolving a specific amount of polymer into a suitable solvent containing mixtures of formic/acetic acid. The ES process involves applying the polymer solution to positively charged electrodes in an electrostatic field, which causes the polymer solution to draw out and spin into random and continues NFs, which are deposited onto a spun-bonded PLA substrate which is rested on a negatively charged collector plate. Area weights were determined by weighing 100 cm^2 samples using a Precisa XB220A analytical balance and by dividing the sample mass by the sample area.

The samples of NF were analysed using a scanning electron microscope JEOL JCM-5000. The Fibraquant evaluation software was used to measure average NF diameters from 50 to 100 measurements.

2.3. Pressure Drop and Breathing Resistance

The total difference in pressure between two points of a fluid (air) as it flows through the filter media is known as the pressure drop. This drop in pressure is caused by frictional forces which are resisting the flow. The drop in pressure is measured by using TexTest FX 3300 LabAir IV, which complies with EN 14683:2019 + AC:2019 [47].

Breathing resistance or breathability measures the difficulty in inhaling or exhaling through a mask or filter media. This is commonly expressed as a pressure drop across the filter media. PALAS PMFT 1000 in accordance with the ASTM Test Method F3502 [41] and NIOSH 42CFR84 (N95) [42] was used to measure different filter media for breathability.

2.4. Filtration Performance Testing

Particle Filtration Efficiency: This procedure was performed to assess the particle filtration efficiency (PFE) of the filter media. Particles of monodispersed polystyrene latex spheres (PSL) were dried, nebulized (atomized), and passed through the filter media and enumerated using a laser particle counter. A one-minute count was performed with the filter media in the system. A control count was performed for one minute without the filter media in the system, before (upstream) and after (downstream) each test article. The filtration efficiency was calculated using the number of particles penetrating the filter media compared to the average of the control values. The air flow rate was maintained at 1 cubic foot per minute (CFM) ±5% during the testing. The procedure used the ASTM F2299 method [40], with some exceptions. While in real world, particles carry a charge, this procedure was carried out in a non-neutralized environment, thus representing a more natural state. The non-neutralized aerosol is also specified in the FDA guidance document on surgical face masks. The % filter penetration (P_{filter}) is calculated as the ratio of the particle concentration downstream (C_{down}) to the particle concentration upstream (C_{up}). The ratio between the particle concentration downstream and the particle concentration upstream is the filter penetration (P_{filter} (%) = $C_{down}/C_{up} \times 100$%). PFE is the complement of the filter penetration (PFE (%) = 100% − P_{filter}). A PFE of 98% means that the filter media will block 98% of particles (either all particle sizes or specific particle sizes) such that only 2% of particles will pass through the material when air is inhaled or exhaled. The filter media were tested in-house and certified by Nelson Lab, ASTM F2299 [40].

Bacterial Filtration Efficiency (BFE): The bacterial filtration efficiency of the material is determined by carrying out a BFE test, which is performed by comparing the bacterial control count upstream of the filter to the bacterial count downstream. A suspension of *Staphylococcus aureus* was aerosolized by a nebulizer and delivered to the test material at a constant rate of flow and air pressure. The challenge delivery was maintained at $1.7\text{–}3.0 \times 10^3$ colony forming units (CFU) with a mean particle size (MPS) of 3.0 ± 0.3 µm. The aerosols were drawn through a six-stage, viable particle Andersen sampler for collection. This test method complies with ASTM F2101-19 [48] and EN 14683:2019 + AC:2019 [47] standards. PLA filter media were tested and certified by Nelson Lab US.

Viral Filtration Efficiency (VFE): The filtration efficiency of the material was determined by performing the VFE test, which is performed by comparing the viral control counts upstream of the filter to the counts downstream.

A suspension of bacteriophage ΦX174 was aerosolized by a nebulizer and delivered to the test material at a constant rate of flow and air pressure. The challenge delivery was maintained at $1.1\text{–}3.3 \times 10^3$ plaque-forming units (PFU) with a mean particle size (MPS) of 3.0 ± 0.3 µm. Then, the aerosol droplets were drawn through a six-stage, viable particle Andersen sampler for collection. The VFE test procedure was adapted from ASTM F2101 standard. PLA filter media were tested and certified by Nelson Lab US.

The PFE values presented in this paper were measured on the filter media only and non-fabricated masks. The PFE of facemasks can differ during filter testing due to air leakages around the edges of the mask. The PLA NF present within the filter media containing an average areal weight of 2.0 ± 0.1 gsm were tested in-house using a PALAS PMFT 1000 testing system according to NIOSH 42CFR84 (N95) [42], ASTM Test Method F3502 and ASTM Test Method F2299 standards [40]. PLA NF filter media were tested and certified by Nelson Lab to pass NIOSH 42CFR84 (N95) [42], ASTM Test Method F3502 [41] and ASTM Test Method F2299 standards [40]. A comparative summary of the filtration test method requirements for the various international test standards are provided in Table 1.

Table 1. Filtration test method requirements.

Test		Level 1	Level 2	Level 3
ASTM F2299 (PFE) Filtration at 0.1 µm—28.3 L·min^{-1}		95%≤	98%≤	
EN14683 (Breathing resistance—Breathability)		≤49 Pa	≤58.8 Pa	
ASTM F2101 -19 (BFE) Filtration at 3 µm—28.3 L·min^{-1}		95%≤	98%≤	
ASTM F2101 (VFE) Filtration at 3 µm—28.3 L·min^{-1}		95%≤	98%≤	
ASTM F3502 Filtration at 0.3 µm—60 L/min		20% ≤	50%≤	-
ASTM F3502 (Breathing resistance—Breathability)		15 mmH$_2$O (147.5 Pa)	5 mmH$_2$O (49 Pa)	-
NIOSH 42 CFR 84 Filtration at 0.3 µm—85 L·min^{-1}			N95 95%≤	
Breathing resistance	Inhalation—120 L·min^{-1}	<314 Pa		ΔP < 98 Pa
	Exhalation—85 L·min^{-1}	<245 Pa		

2.5. Electrospinning Solutions

Electrospinning solutions were prepared by stirring PLA polymer in a mixture of formic acid/acetic acid solvent at room temperature until all the polymers had fully dissolved. Manuka triketone extracted from manuka oil was then added to the solution.

2.6. Antimicrobial Activity

FilterLayr Eco consisting of a functional nanofiber layer sandwiched between two layers of PLA spunbonded nonwoven fabric was tested for antimicrobial activity. The tests followed the protocols outlined in ISO 18184:2019 [44], Determining virucidal activity of textile product, ISO 20743—Determination of antibacterial activity of textile product [43] and ASTM G21—Resistance of Synthetic Polymeric Materials to Fungi [45]. The antimicrobial testing protocol can be found in Figures S1 and S2 in the supporting information document.

2.6.1. Antibacterial Activity

The PLA filter media were tested against bacterium *Staphylococcus aureus* (ATCC 6538P *Escherichia coli* (ATCC 8739) and *Klebsiella pneumoniae* (ATCC 4352) in accordance with the following test standard ISO 20743:2013—Quantitative antibacterial test on textiles [43]. In addition, the PLA filter media were subjected to 5 and 10 laundering cycles following ISO 6330:2013—Domestic laundering [49], and then tested against Gram-positive bacterium *Staphylococcus aureus* and Gram-negative bacterium *Klebsiella pneumoniae* following ISO 20743:2013—Quantitative antibacterial test method standards [43]. The antibacterial activity rating is shown in Table 2.

Table 2. Antibacterial activity rating associated with percentage reduction of bacterial growth.

Antibacterial Activity Value (A)	Efficacy Rating	A Value	Reduction (%)
$2 \leq A < 3$	Good Effect Level	1	90
		2	99
		3	99.9
$A \geq 3$	Excellent Effect Level	4	99.99
		5	99.999

Antibacterial Activity Calculation:

$$A = (C_t - C_0) - (T_t - C_0) = (C_t - T_t) \tag{1}$$

where A is the antibacterial activity value, C_0 is the logarithm average of 3 bacterial colony forming units (cfu) immediately after inoculation of the control specimen, C_t is the logarithm average of 3 bacterial colony-forming units (cfu) after specified contact time with the control specimen, and T_t is the logarithm average of 3 bacterial colony-forming units (cfu) after specified contact time with the treated specimen.

Percent reduction calculation:

$$\text{Percent Reduction} = 1 - 10^A \tag{2}$$

2.6.2. Antiviral Activity

The PLA filter media were tested against three highly potent and prevalent viruses such as Influenza A (H1N1) (ATCC VR-1469), human coronavirus 229E (ATCC VR-740) and SARS-CoV-2 (Delta variant, B.1.617.2; NCBI MZ574052) in accordance with the method outlined in ISO 18184:2019 standard [44]. The antiviral activity rating is shown in Table 3.

Table 3. Antiviral activity rating associated with percentage reduction of virus growth.

Antiviral Activity Value (Mv)	Efficacy Rating	Mv Value	Reduction (%)
		1	90
$2 \leq M_v < 3$	Good Effect Level	2	99
		3	99.9
$M_v \geq 3$	Excellent Effect Level	4	99.9
		5	99.999

Anti-viral activity calculation:

$$M_v = \log_{10}(V_a) - \log_{10}(V_b) \qquad (3)$$

where M_v is the antiviral activity value, $\log_{10}(V_a)$ is the logarithm average of 3 infectivity titre values immediately after inoculation of the control specimen, and $\log_{10}(V_b)$ is the logarithm average of 3 infectivity titre values after specified contact time with the control specimen.

Percent reduction calculation:

$$\text{Percent Reduction} = 1 - 10^{-M_v} \qquad (4)$$

2.6.3. Anti-Fungal Activity

Control and test specimen of 2" × 2" dimension was cut for the testing. Fungal species were grown separately on Sabouraud dextrose agar for 7–14 days. The spore suspension of the fungi was prepared by pouring 10 mL of sterile DI water containing 0.5 mL of Tween 20 into the culture plate. The surface growth was gently scraped from the culture. The spore suspension was transferred into a centrifuge tube containing 25 mL of sterile DI water. The centrifuge tube was vortexed for one minute to break the spore clumps. The spore suspension was filtered to remove mycelial fragments. The spore suspension was washed three times in DI water by centrifugation and diluted to achieve a 1.0×10^6 spore/mL for each fungal species. Spore suspensions were then combined using equal volumes of resultant spore suspension. Both test sample and control were placed separately onto Sabouraud dextrose agar, and an even layer of spore suspension was sprayed onto each material sample. Plates were incubated at 29 °C ± 1 °C and examined weekly for 28 days (Relative humidity >85%). All tests were performed in triplicate. The PLA filter media were tested against *Aspergillus niger* (AATCC 16888) following ASTM G21 standard [45].

2.7. Washability of the Filter Media

Laundering of the PLA NF filter media were performed in accordance with ISO 6330 [49] standard using the following test parameters: 3G:30 °C, gentle setting (wool/silk/synthetics). The washed PLA NFs filter media were tested for filtration performance following ASTM F3502 [41] protocol and tested against *Staphylococcus aureus* and *Klebsiella pneumoniae* following test method outlined in ISO 20743 standard [43].

3. Results and Discussion

NanoLayr Ltd., based in Auckland, New Zealand, is a producer of large-scale advanced nanofiber textiles certified AS9100d. We have recently launched a unique filter media marketed as FilterLayr™ Eco. The product has a three-layer structure which consists of a middle layer composed of PLA NFs and manuka triketone at an average areal weight of 2.0 ± 0.1 gsm sandwiched between two outer layers of PLA spun-bonded non-woven fabric (Figure 1). Figure 1B,C displays SEM images of NFs produced from PLA solutions, randomly oriented and with an average fibre diameter of 168.3 ± 43.6 nm. The NFs display the typical stacked layered morphology of non-woven fibrous fabrics. The implementation of nano-scale fibres has been shown to be advantageous in air filtration due to the small diameter of the fibres and the corresponding high surface-to-volume ratio, enhancing the capture of particles through interception and other mechanisms [11,19,50,51]. Furthermore, homogeneous porosity can lead to lower pressure drop due to slip flow effects, made possible by the small fibre diameter when compared to microfiber counterparts [50].

Figure 1. (**A**) Diagram of FilterLayr^{TM} Eco structure spun-bond PLA/PLA electrospun nanofibre/spun-bond PLA; (**B**) Scanning electron micrograph of nanofibre layer made from PLA; and (**C**) average fibre distribution of PLA electrospun fibres using Fibraquant image analysis software.

3.1. Filtration Performance of PLA NF Filter Media

In a previous work, the degree of NF uniformity has been demonstrated by plotting the pressure drop of the NF filter media against the areal weight of NF material, showing linearity [11]. In an idealised situation, an increase in pressure drop will result in a proportional increase in filtration efficiency of the material. Figure 2 shows the relationship between the filtration efficiency and the pressure drop of the filter media when tested at three different air velocities as determined by international test standards (A) ASTM F2299 [40], (B) NIOSH 42CFR84 [42], and (C) ASTM F3502 [41]. All the filter media showed a linear relationship between pressure drop and filtration efficiency, with R^2 values ranging from 0.848 to 0.979 depending on the test standard.

Up to 13 individual samples of the PLA NF filter media were tested according to the three international standards. Filtration efficiency and pressure drop values are displayed in Table 4. Filter media samples were challenged with monodispersed polystyrene latex sphere (PSL) aerosols with an average particle diameter of 0.1 μm (Figure 2A) and with NaCl aerosol with an average particle diameter of 0.3 μm (Figure 2B,C). The PLSs were nebulised, dried, and passed through the filter media at an airflow velocity of 28.3 L·min^{-1}, as required for the PFE test method in ASTM F2299 [40]. All filter media samples containing PLA NF layers met the particulate filtration requirement for level 2 as specified in ASTM F2299 standard [40]. It displayed filtration efficiencies ≥98% (Figure 2A) and relatively low pressure drops ranging from 48 to 59 Pa (air velocities = 8 L·min^{-1}). In addition, samples of the same areal weight were sent to a third-party-certified laboratory to ensure the efficacy of PLA NF filter media to pass the requirements of the test method ASTM 2100 [51]. The PFE, VFE and BFE results are presented in the supporting document (Tables S1–S5). The filter media exceeded the level 2 criteria outlined in the ASTM F2100 standard [51] while following the efficiency values obtained for the various tests, PFE results above 99.5% (Table S1), VFE results above 99.9% (Table S1), and BFE results above 99% (Table S4), with an airflow resistance requirement below 58.8 Pa (Table S3).

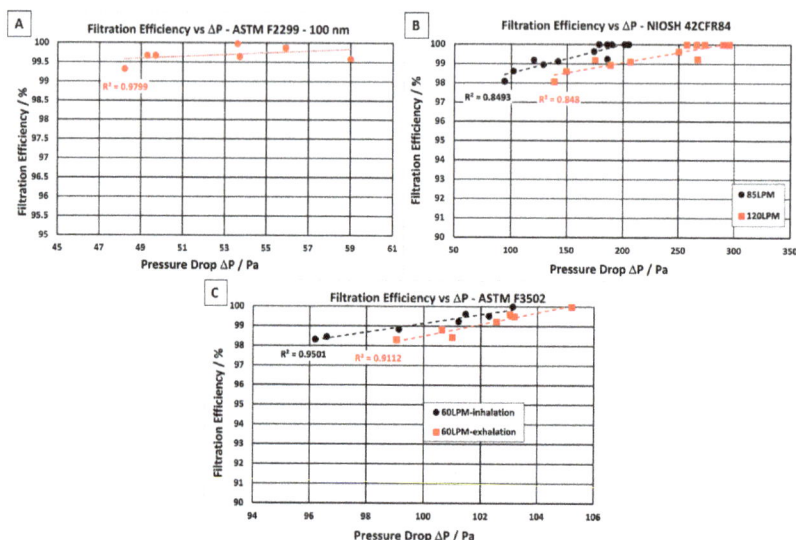

Figure 2. Filtration efficiency vs. pressure drop for PLA NF filter media tested in accordance with the following international standards: (**A**) ASTM F2299 [40], (**B**) NIOSH 42CFR84 [42] and (**C**) ASTM F3502 [41].

Table 4. Pressure drop and filtration efficiency results for PLA NF filter media tested in accordance with ASTM F2299 [40], ASTM F3502 [41] and NIOSH 42CFR84 [42].

	ASTM F2299		ASTM F3502			NIOSH 42CFR84		
	Filtration Efficiency at 100 nm	ΔP	Filtration Efficiency at 300 nm	ΔP Inh. at 60 LPM	ΔP Exh. at 60 LPM	Filtration Efficiency at 300 nm	ΔP Ex. at 85 LPM	ΔP Inh. at 120 LPM
1	99.87	55.90	98.31	96.20	99.05	98.08	94.20	138.58
2	99.88	55.90	98.44	96.60	100.99	98.61	102.00	149.37
3	99.59	59.00	98.83	99.13	100.63	99.18	120.47	174.84
4	99.65	53.70	99.23	101.21	102.54	98.95	129.13	188.36
5	99.68	49.30	99.62	101.46	103.01	99.11	142.08	206.31
6	99.33	48.20	99.51	102.28	103.16	99.64	174.00	250
7	99.98	53.60	100	103.12	105.21	100	178.44	257.3
8	99.68	49.70				100	185.48	266.18
9						99.24	185.77	266.95
10						100	189.76	273.38
11						100	201.14	289.63
12						100	204.18	295.12
13						100	205.14	295.33

The PLA filter media were also challenged with NaCl aerosol with a count median diameter (CMD) of 75 ± 20 nm and a geometric standard deviation (GSD) of 1.86, equivalent to an average particle diameter of 0.3 µm (Figure 2B,C) which is used for both NIOSH 42CFR84 N95 [42] and ASTM F3502 test methods [41]. The NaCl particles were nebulised, dried, and passed through the filter media at an airflow velocity of 60 or 85 L·min^{-1} as required for the PFE test method in ASTM F3502 [41], and NIOSH 42CFR84 N95 [42], respectively. The PLA filter media exceeded the NIOSH 42CFR84 N95 particle filtration efficiency requirements of ≥95% as well as meeting the N95 inhalation and exhalation requirements of <314 and <245 Pa at 85 and 120 L·min^{-1}, respectively.

In the standard for barrier face coverings, ASTM F3502 [41], a minimum filtration efficiency of 20% is necessary to meet level 1, and ≥50% filtration efficiency is required to meet level 2. All PLA NF filter media samples tested exceeded the filtration performance requirements of ASTM F3502 level 2 [41]. Up to 98% filtration efficiency was achieved by the PLA NF filter media. Each of the samples tested in accordance with ASTM F3502 passed the level 1 breathability requirements. The PLA filter media were also tested in a certified laboratory to ensure their ability to pass the requirements of ASTM F3502 [41] and NIOSH 42CFR84 N95 [42]. The reports from the certified laboratory are presented in the supporting document (Tables S6–S8).

To verify the reusability of the filter media, the samples were subjected to 10 laundering cycles and then challenged with NaCl particles, as described above. The results showed a slight decrease in filtration efficiency (Table S7). Out of the 10 samples that were laundered, one was damaged during the laundering cycles, and therefore was not tested. The remaining nine samples that were tested achieved an average filtration efficiency of 58.97% which readily meets the minimum requirement for level 2 filtration efficiency, in accordance with ASTM F3502 standard [41]. One of the nine samples that were tested is also suspected to have suffered some damage during laundering which consequently affected the overall average result.

3.2. Antimicrobial Activity of PLA Nanofibers Containing Manuka Triketone

In this work, natural and non-toxic manuka oil 5% wt. has been used to generate antimicrobial nanofibre filter media. The antibacterial properties of the filter media were tested according to ISO 20743:2013 [43] standard using three different bacteria: ATCC 6538P (*Staphylococcus aureus*), ATCC 4352 (*Klebsiella pneumoniae*) and ATCC 8739 (*Escherichia coli*) (Figure 3, left and Table 5). The electrospun PLA filter media were challenged against the test standard ISO 18184:2019 [44] using three different viruses: *influenza A* [H1N1] (ATCC VR-1469), human coronavirus 229E (ATCC VR-740) and SARS-CoV-2 (Delta variant B.1.617.2; NCBI MZ574052) (Figure 3, right and Table 6). Additionally, antifungal activity was tested according to ASTM G21 using ATCC 16888 (*Aspergillus niger*) (Table 7) [45].

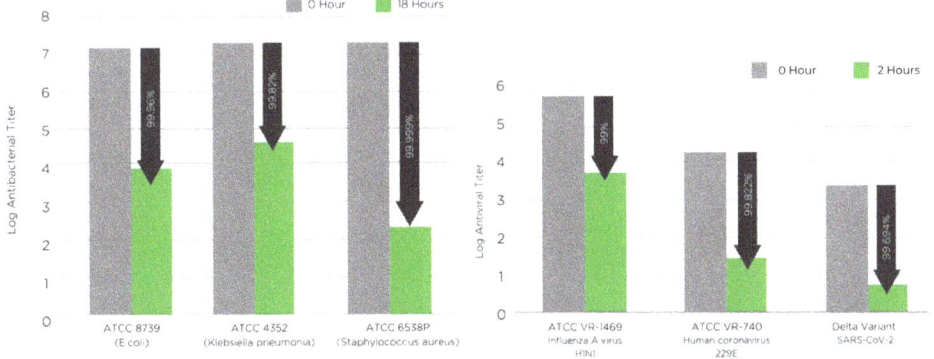

Figure 3. (**left**) Antibacterial activity of PLA NF filter media containing manuka triketone (5 wt.%) against *S. aureus* (ATCC 6538P), *E. coli* (ATCC 8739) and *K. pneumoniae* (ATCC 4352) tested according to ISO 20743:2013 [43]. (**right**) Antiviral activity of PLA NF filter media containing manuka triketone (5 wt.%) against influenza A, human coronavirus 229E and SARS-CoV-2 tested according to ISO 18184:2019 [44]. The arrows point at the inhibition of microbial growth.

Table 5. Antibacterial reduction rate of PLA NF filter media containing manuka triketone (5 wt.%) against *S. aureus* (ATCC 6538P), *E. coli* (ATCC 8739). and *K. pneumoniae* (ATCC 4352) tested according to ISO 20743:2013 [43]. Reusability following 5 and 10 laundering cycle according to ISO 6330:2013 [49] and ISO 20743:2013, was confirmed for the PLA NF filter media when tested against *S. aureus* (ATCC 6538P) and *K. pneumoniae* (ATCC 4352).

ISO 20743:2013	S. aureus (ATCC 6538P)			E. Coli (ATCC 8739)	K. pneumonia (ATCC 4352)		
	0 Wash	5 Washes	10 Washes	0 Wash	0 Wash	5 Washes	10 Washes
Control	Initial			Initial	Initial		
Log CFU	4.69			4.69	4.83	4.80	
Contacting time (hours)	-			-	-	-	
Control	After contacting			After contacting	After contacting	After contacting	
Log CFU	7.25			7.19	7.3	7.6	
Contacting time (hours)	18			18	18	18	
Log CFU samples	2.33	2.87	3.1	3.83	4.55	4.82	5.11
Percentage reduction samples (%)	99.999	99.996	99.993	99.957	99.82	99.834	99.678
Log reduction samples	4.92	4.38	4.15	3.36	2.75	2.78	2.49

Table 6. Antiviral reduction rate of PLA NF filter media containing manuka triketone (5 wt.%) against influenza A, human coronavirus 229E and SARS-CoV-2 tested according to ISO 18184:2019 [44].

ISO 18184:2019	Influenza A (H1N1) (ATCC VR-1469)	Human Coronavirus 229E (ATCC VR-740)	SARS-CoV-2 (Delta Variant)
Infective titer test	TCID50 method	TCID50 method	Plaque assay
Log(Va) (Control, immediately)	5.88	4.25	3.21
Contacting time (hours)	2	2	2
Log(Vc) (Sample, after contacting)	3.88	1.5	0.70
Antiviral activity value, Mv	2	2.8	2.5
Percentage reduction samples (%)	99	99.82	99.69

Table 7. Antifungal activity of PLA NF filter media containing manuka triketone (5 wt.%) against *Aspergillus niger* tested according to ASTM G21 [45]. Note on antifungal rating: 0 no growth, 1 trace of growth (less than 10% coverage), 2 light growth (10–30% coverage), 3 medium growth (30–60% coverage) and 4 heavy growth (60–100% coverage).

Sub-Samples	Control Specimen	Test Specimen
Contact time	28 days	28 days
Rating	4	0
Observed Growth	High Growth	No Growth

3.2.1. Antibacterial Activity of PLA Nanofibers Containing Manuka Triketone

The antibacterial activity of the filter media were tested according to ISO 20743:2013 [43]. The antibacterial test results confirmed that the manuka oil was integrated into the PLA NF mat and imparted excellent biocidal action against both Gram-negative and Gram-positive bacteria. The nanofibers loaded with manuka oil showed bacterial population reductions of 99.999% against *S. aureus*, 99.96% against *E. coli*. and 99.82% against *K. pneumoniae strains*, after a contact time of 18 h (Figure 3 and Table 5).

In order to test the retention of antimicrobial activity of the filter media after multiple washes, the samples were subjected to 5 and 10 laundering cycles, and the antibacterial performance was re-evaluated (Table 5). The laundering step was carried out in accordance with ISO 6330:2013 [49] standard using the following test parameters: 3G:30 °C, gentle setting (wool/silk/synthetics). Both the Gram-positive (ATCC 6538P) and Gram-negative (AATCC 4352) bacteria were used to challenge the PLA NF filter media. The sample showed excellent antibacterial activity even after 5 and 10 laundering cycles. The reduction rates of *Staphylococcus aureus* and *Klebsiella pneumoniae* bacteria after five washes upon a contact time of 18 h were 99.996% and 99.834%, respectively. After 10 washes, the reduction rates of *Staphylococcus aureus* and *Klebsiella pneumoniae* after a contact time of 18 h were 99.993% and 99.678%, respectively.

3.2.2. Antiviral Activity of PLA Nanofibers Containing Manuka Triketone

The antiviral results of the filter media were tested according to ISO 18184:2019 [44]. The results of the tests showed that the manuka oil embedded into the PLA nanofibers has outstanding virucidal properties against influenza A (H1N1), human coronavirus 229E and SARS-CoV-2 (Delta variant) (Figure 3, right and Table 6). The PLA filter media showed a virus reduction of 99% against influenza A, 99.82% against human coronavirus 229E and 99.69% against SARS-CoV-2 after a contact time of 2 h.

3.2.3. Antifungal Activity of PLA Nanofibers Containing Manuka Triketone

The antifungal/fungicidal activity of the PLA NF filter media were evaluated using the following test method: ASTM G21—Resistance of Synthetic Polymeric Materials to Fungi (Table 7) [45]. The filter media displayed antifungal property and showed no *Aspergillus niger* growth after a period of 28 days (Table 7).

4. Conclusions

Novel antimicrobial air filtration media (marketed as FilterLayr^{TM} Eco) containing electrospun PLA NFs loaded with manuka triketone have been developed by NanoLayr Ltd. FilterLayr^{TM} Eco has been proven to have excellent particle filtration efficiency and breathability and meets the requirements for following standard certifications, NIOSH N95, ASTM F2100 level 2, and ASTM F3502 level 1 breathing resistance and level 2 filtration efficiency. In addition to this, the filter media successfully passed ASTM F3502 level 2 filtration efficiency after 10 laundering cycles, making it reusable for a minimum of 10 washes.

FilterLayr^{TM} Eco is unique in its ability to both trap and kill airborne bacteria and viruses using naturally occurring antimicrobial Manuka oil as the biocidal agent. This enables the filter media to self-decontaminate and allows it to be used multiple times before it needs to be replaced, as it does not harbour or proliferate the microbes that it captures. This study demonstrated the effectiveness of the PLA filter media in neutralising different types of bacteria, viruses, and a fungus, namely, *Staphylococcus aureus*, *Klebsiella pneumoniae*, *Escherichia coli*, influenza A, human coronavirus 229E, SARS-CoV-2-Delta variant and *Aspergillus niger*. The filter media has also shown excellent performance against bacteria when tested after 5 and 10 laundering cycles, thus proving its ability to retain its anti-microbial properties for a minimum of 10 wash cycles, making it reusable and more sustainable. FilterLayr^{TM} Eco is also made from industrially compostable polymers, thus reducing the amount of waste being sent to landfills and consequently reducing plastic waste in the environment.

Supplementary Materials: The following supporting information can be downloaded at: https://www.mdpi.com/article/10.3390/membranes12060571/s1, Figure S1. Illustration of the Antibacterial testing procedure, Figure S2. Illustration of the Antiviral testing procedure, Table S1. Filtration efficiency of PLA NF filter when challenge with 0.1 μm PSL particles following ASTM F2299 test method, certified by Nelson Lab, Table S2. Particles filtration efficiency of PLA NF filter when challenge with 0.3 μm PSL particles following ASTM F2299 test method, certified by Nelson Lab, Table S3. Pressure drops values of PLA NF filter when tested against EN 14683:2019+AC:2019 test method, certified by Nelson Lab, Table S4. Bacterial filtration efficiency of PLA NF filter media, Table S5. Viral filtration efficiency of PLA NF filter media, Table S6. Filtration efficiency of PLA NF filter when challenge with 0.3 μm NaCl particles fol-lowing ASTM F3502 test method, certified by Nelson Lab, Table S7. Filtration efficiency of PLA NF filter when challenge with 0.3 μm NaCl particles fol-lowing ASTM F3502 test method after 10 laundering cycles, certified by Nelson Lab, Table S8. Filtration efficiency of PLA NF filter when challenge with 0.3 μm NaCl particles fol-lowing NIOSH—42CFR84 test method, certified by Nelson Lab.

Author Contributions: Writing—original draft preparation, F.N.H.K. and D.F.; writing—review and editing, F.N.H.K. and D.F.; supervision, G.W.B. methodology, N.A.C. and G.H. All authors have read and agreed to the published version of the manuscript.

Funding: This research received no external funding.

Institutional Review Board Statement: Not applicable.

Informed Consent Statement: Not applicable.

Data Availability Statement: The data presented in this study are available on request from the corresponding author.

Conflicts of Interest: The authors declare no conflict of interest.

References

1. Babaahmadi, V.; Amid, H.; Naeimirad, M.; Ramakrishna, S. Biodegradable and multifunctional surgical face masks: A brief review on demands during COVID-19 pandemic, recent developments, and future perspectives. *Sci. Total Environ.* **2021**, *798*, 149233. [CrossRef] [PubMed]
2. Huang, H.; Fan, C.; Li, M.; Nie, H.-L.; Wang, F.-B.; Wang, H.; Wang, R.; Xia, J.; Zheng, X.; Zuo, X.; et al. COVID-19: A Call for Physical Scientists and Engineers. *ACS Nano* **2020**, *14*, 3747–3754. [CrossRef] [PubMed]
3. Pacitto, A.; Amato, F.; Salmatonidis, A.; Moreno, T.; Alastuey, A.; Reche, C.; Buonanno, G.; Benito, C.; Querol, X. Effectiveness of commercial face masks to reduce personal PM exposure. *Sci. Total Environ.* **2019**, *650*, 1582–1590. [CrossRef] [PubMed]
4. Prather Kimberly, A.; Wang Chia, C.; Schooley Robert, T. Reducing transmission of SARS-CoV-2. *Science* **2020**, *368*, 1422–1424. [CrossRef]
5. Pullangott, G.; Kannan, U.; Gayathri, S.; Kiran, D.V.; Maliyekkal, S.M. A comprehensive review on antimicrobial face masks: An emerging weapon in fighting pandemics. *RSC Adv.* **2021**, *11*, 6544–6576. [CrossRef]
6. Pasanen, A.L.; Keinanen, J.; Kalliokoski, P.; Martikainen, P.I.; Ruuskanen, J. Microbial growth on respirator filters from improper storage. *Scand. J. Work. Environ. Health* **1993**, *6*, 421–425. [CrossRef]
7. Brosseau, L.M.; McCullough, N.V.; Vesley, D. Bacterial Survival on Respirator Filters and Surgical Masks. *J. Am. Biol. Saf. Assoc.* **1997**, *2*, 32–43. [CrossRef]
8. Zheng, C.R.; Li, S.; Ye, C.; Li, X.; Zhang, C.; Yu, X. Particulate Respirators Functionalized with Silver Nanoparticles Showed Excellent Real-Time Antimicrobial Effects against Pathogens. *Environ. Sci. Technol.* **2016**, *50*, 7144–7151. [CrossRef]
9. Fadare, O.O.; Okoffo, E.D. Covid-19 face masks: A potential source of microplastic fibers in the environment. *Sci. Total Environ.* **2020**, *737*, 140279. [CrossRef]
10. Rengasamy, S.; Fisher, E.; Shaffer, R.E. Evaluation of the survivability of MS2 viral aerosols deposited on filtering face piece respirator samples incorporating antimicrobial technologies. *Am. J. Infect. Control.* **2010**, *38*, 9–17. [CrossRef]
11. Karabulut, F.N.H.; Höfler, G.; Ashok Chand, N.; Beckermann, G.W. Electrospun Nanofibre Filtration Media to Protect against Biological or Nonbiological Airborne Particles. *Polymers* **2021**, *13*, 3257. [CrossRef] [PubMed]
12. Essa, W.K.; Yasin, S.A.; Saeed, I.A.; Ali, G.A.M. Nanofiber-Based Face Masks and Respirators as COVID-19 Protection: A Review. *Membranes* **2021**, *11*, 250. [CrossRef] [PubMed]
13. Zhang, Z.; Ji, D.; He, H.; Ramakrishna, S. Electrospun ultrafine fibers for advanced face masks. *Mater. Sci. Eng. R Rep.* **2021**, *143*, 100594. [CrossRef] [PubMed]
14. Ramakrishna, S.; Fujihara, K.; Teo, W.E.; Yong, T.; Ma, Z.; Ramaseshan, R. Electrospun nanofibers; solving global issues. *Mater. Today* **2006**, *9*, 40–50. [CrossRef]

15. Prata, J.C.; Silva, A.L.P.; Walker, T.R.; Duarte, A.C.; Rocha-Santos, T. COVID-19 Pandemic Repercussions on the Use and Management of Plastics. *Environ. Sci. Technol.* **2020**, *54*, 7760–7765. [CrossRef]
16. Hamdan, N.; Yamin, A.; Hamid, S.A.; Khodir, W.K.; Guarino, V. Functionalized Antimicrobial Nanofibers: Design Criteria and Recent Advances. *J. Funct. Biomater.* **2021**, *12*, 59. [CrossRef]
17. Gao, Y.; Bach Truong, Y.; Zhu, Y.; Louis Kyratzis, I. Electrospun antibacterial nanofibers: Production, activity, and in vivo applications. *J. Appl. Polym. Sci.* **2014**, *131*, 40797. [CrossRef]
18. Kumar, S.; Jang, J.; Oh, H.; Jung, B.J.; Lee, Y.; Park, H.; Yang, K.H.; Chang Seong, Y.; Lee, J.-S. Antibacterial Polymeric Nanofibers from Zwitterionic Terpolymers by Electrospinning for Air Filtration. *ACS Appl. Nano Mater.* **2021**, *4*, 2375–2385. [CrossRef]
19. Pardo-Figuerez, M.; Chiva-Flor, A.; Figueroa-Lopez, K.; Prieto, C.; Lagaron, J.M. Antimicrobial Nanofiber Based Filters for High Filtration Efficiency Respirators. *Nanomaterials* **2021**, *11*, 900. [CrossRef]
20. Song, K.; Wu, Q.; Zhang, Z.; Ren, S.; Lei, T.; Negulescu, I.I.; Zhang, Q. Porous Carbon Nanofibers from Electrospun Biomass Tar/Polyacrylonitrile/Silver Hybrids as Antimicrobial Materials. *ACS Appl. Mater. Interfaces* **2015**, *7*, 15108–15116. [CrossRef]
21. Castro-Mayorga, J.L.; Fabra, M.J.; Lagaron, J.M. Stabilized nanosilver based antimicrobial poly(3-hydroxybutyrate-co-3-hydroxyvalerate) nanocomposites of interest in active food packaging. *Innov. Food Sci. Emerg. Technol.* **2016**, *33*, 524–533. [CrossRef]
22. Blosi, M.; Costa, A.L.; Ortelli, S.; Belosi, F.; Ravegnani, F.; Varesano, A.; Tonetti, C.; Zanoni, I.; Vineis, C. Polyvinyl alcohol/silver electrospun nanofibers: Biocidal filter media capturing virus-size particles. *J. Appl. Polym. Sci.* **2021**, *138*, 51380. [CrossRef] [PubMed]
23. Ditaranto, N.; Basoli, F.; Trombetta, M.; Cioffi, N.; Rainer, A. Electrospun Nanomaterials Implementing Antibacterial Inorganic Nanophases. *Appl. Sci.* **2018**, *8*, 1643. [CrossRef]
24. Zhang, W.; Ronca, S.; Mele, E. Electrospun Nanofibres Containing Antimicrobial Plant Extracts. *Nanomaterials* **2017**, *7*, 42. [CrossRef]
25. Yao, C.H.; Yeh, J.Y.; Chen, Y.S.; Li, M.H.; Huang, C.H. Wound-healing effect of electrospun gelatin nanofibres containing Centella asiatica extract in a rat model. *J. Tissue Eng. Regen. Med.* **2017**, *11*, 905–915. [CrossRef]
26. Jin, G.; Prabhakaran, M.P.; Kai, D.; Annamalai, S.K.; Arunachalam, K.D.; Ramakrishna, S. Tissue engineered plant extracts as nanofibrous wound dressing. *Biomaterials* **2013**, *34*, 724–734. [CrossRef]
27. Agnes Mary, S.; Giri Dev, V.R. Electrospun herbal nanofibrous wound dressings for skin tissue engineering. *J. Text. Inst.* **2015**, *106*, 886–895. [CrossRef]
28. Lin, S.; Chen, M.; Jiang, H.; Fan, L.; Sun, B.; Yu, F.; Yang, X.; Lou, X.; He, C.; Wang, H. Green electrospun grape seed extract-loaded silk fibroin nanofibrous mats with excellent cytocompatibility and antioxidant effect. *Colloids Surf. B Biointerfaces* **2016**, *139*, 156–163. [CrossRef]
29. Patel, B. Corporate Uniform Fabrics with Antimicrobial Edge: Preparation and Evaluation Methodology. *Int. Dye.* **2014**, *2*, 33–38.
30. Radetić, M.; Marković, D. Nano-finishing of cellulose textile materials with copper and copper oxide nanoparticles. *Cellulose* **2019**, *26*, 8971–8991. [CrossRef]
31. Rezaie, A.B.; Montazer, M.; Rad, M.M. A cleaner route for nanocolouration of wool fabric via green assembling of cupric oxide nanoparticles along with antibacterial and UV protection properties. *J. Clean. Prod.* **2017**, *166*, 221–231. [CrossRef]
32. Kale, R.; Vade, A.; Kane, P. Antibacterial and Conductive Polyester Developed using Nano Copper Oxide and Polypyrrole Coating. *J. Text. Assoc.* **2019**, *79*, 340–346.
33. Nischala, K.; Rao, T.N.; Hebalkar, N. Silica–silver core–shell particles for antibacterial textile application. *Colloids Surf. B Biointerfaces* **2011**, *82*, 203–208. [CrossRef] [PubMed]
34. Eremenko, A.M.; Petrik, I.S.; Smirnova, N.P.; Rudenko, A.V.; Marikvas, Y.S. Antibacterial and Antimycotic Activity of Cotton Fabrics, Impregnated with Silver and Binary Silver/Copper Nanoparticles. *Nanoscale Res. Lett.* **2016**, *11*, 28. [CrossRef] [PubMed]
35. Ibrahim, N.A.; Eid, B.M.; Hashem, M.M.; Refai, R.; El-Hossamy, M. Smart Options for Functional Finishing of Linen-containing Fabrics. *J. Ind. Text.* **2010**, *39*, 233–265. [CrossRef]
36. Irfan, M.; Perero, S.; Miola, M.; Maina, G.; Ferri, A.; Ferraris, M.; Balagna, C. Antimicrobial functionalization of cotton fabric with silver nanoclusters/silica composite coating via RF co-sputtering technique. *Cellulose* **2017**, *24*, 2331–2345. [CrossRef]
37. Merkl, P.; Long, S.; McInerney, G.M.; Sotiriou, G.A. Antiviral Activity of Silver, Copper Oxide and Zinc Oxide Nanoparticle Coatings against SARS-CoV-2. *Nanomaterials* **2021**, *11*, 1312. [CrossRef]
38. Pollard, Z.A.; Karod, M.; Goldfarb, J.L. Metal leaching from antimicrobial cloth face masks intended to slow the spread of COVID-19. *Sci. Rep.* **2021**, *11*, 19216. [CrossRef]
39. Kapoor, N.; Yadav, R. Manuka honey: A promising wound dressing material for the chronic nonhealing discharging wounds: A retrospective study. *Natl. J. Maxillofac. Surg.* **2021**, *12*, 233–237. [CrossRef]
40. ASTM International—ASTM F2299-03—Standard Test Method for Determining the Initial Efficiency of Materials Used in Medical Face Masks to Penetration by Particulates Using Latex Spheres. Available online: https://www.astm.org/f2299-03.html (accessed on 20 April 2022).
41. ASTM International—ASTM F3502—Barrier Face Coverings and Workplace Performance/Performance Plus Masks. Available online: https://www.astm.org/f3502-21.html (accessed on 20 April 2022).
42. NIOSH 42 CFR Part 84—Respiratory Protective Devices. Available online: https://www.cdc.gov/niosh/npptl/topics/respirators/pt84abs2.html (accessed on 20 April 2022).

43. ISO 20743:2013 Textiles—Determination of Antibacterial Activity of Textile Products. Available online: https://www.iso.org/standard/59586.html (accessed on 20 April 2022).
44. ISO 18184:2019 Textiles—Determination of Antiviral Activity of Textile Products. Available online: https://www.iso.org/standard/71292.html (accessed on 20 April 2022).
45. ASTM G21-15(2021)e1—Standard Practice for Determining Resistance of Synthetic Polymeric Materials to Fungi. Available online: https://standards.globalspec.com/std/14393372/astm-g21-15-2021-e1 (accessed on 20 April 2022).
46. Rutledge, G.C.; Fridrikh, S.V. Formation of fibers by electrospinning. *Adv. Drug Deliv. Rev.* **2007**, *59*, 1384–1391. [CrossRef]
47. EN 14683:2019+AC:2019—Medical Face Masks—Requirements and Test Methods. Available online: https://standards.iteh.ai/catalog/standards/cen/13eebf8f-0ff1-4084-946a-2fb8f6e24928/en-14683-2019 (accessed on 20 April 2022).
48. ASTM International—ASTM F2101-19 -Standard Test Method for Evaluating the Bacterial Filtration Efficiency (BFE) of Medical Face Mask Materials, Using a Biological Aerosol of Staphylococcus Aureus. Available online: https://www.astm.org/f2101-19.html (accessed on 20 April 2022).
49. ISO 6330:2021 Textiles—Domestic Washing and Drying Procedures for Textile Testing. Available online: https://www.iso.org/standard/75934.html (accessed on 20 April 2022).
50. Li, L.; Frey, M.W.; Green, T.B. Modification of Air Filter Media with Nylon-6 Nanofibers. *J. Eng. Fibers Fabr.* **2006**, *1*, 155892500600100101. [CrossRef]
51. ASTM International—ASTM F2100-21—Standard Specification for Performance of Materials Used in Medical Face Masks. Available online: https://standards.globalspec.com/std/14468211/astm-f2100-21 (accessed on 20 April 2022).

Article

Polypropylene Hollow-Fiber Membrane Made Using the Dissolution-Induced Pores Method

Zhongyong Qiu * and Chunju He

The State Key Laboratory for Modification of Chemical Fibers and Polymer Materials, College of Materials Science and Engineering, Donghua University, Shanghai 201620, China; chunjuhe@dhu.edu.cn
* Correspondence: 1185080@mail.dhu.edu.cn; Tel.: +86-182-1756-5783

Abstract: The efficient preparation of hydrophilic polypropylene membranes has always been a problem. Here, a twin-screw extruder was used to melt-blend ethylene-vinyl alcohol copolymer and polypropylene; then, hollow fibers were extrusion-molded with a spinneret and taken by a winder; after this, dimethyl sulfoxide was used to dissolve the ethylene-vinyl alcohol copolymer of the fiber to obtain a polypropylene hollow-fiber membrane. This procedure was used to study the effects of different contents and segment structure of ethylene-vinyl alcohol copolymer on the structure and filtration performance of the membranes; furthermore, the embedded factor and blocked factor were used to evaluate the ethylene-vinyl alcohol copolymer embedded in the matrix without dissolving and or being completely blocked in the matrix, respectively. The results show that the increase in ethylene-vinyl alcohol copolymer could reduce the embedded factor and increase the blocked factor. The increase in the polyethylene segments of ethylene-vinyl alcohol copolymer could increase both the embedded factor and blocked factor. The water permeation of the membrane reached 1300 L$m^{-2} \cdot h^{-1} \cdot bar^{-1}$ with a 100% rejection of ink (141 nm) and the elongation at break reached 188%, while the strength reached 22 MPa. The dissolution-induced pores method provides a completely viable alternative route for the preparation of polypropylene membranes.

Keywords: PP membrane; dissolution-induced pore method; polypropylene hollow-fiber membrane

1. Introduction

Polypropylene hollow-fiber membrane (PPHFM) is widely used in water treatment [1–3], industrial purification [4,5], and pharmaceutical separation [6–8]. Currently, polypropylene (PP) hollow-fiber membranes are mainly prepared with the thermally induced phase separation method (TIPS) and the melt-stretching method (MS) [9,10]. The studies on these two methods are comprehensive. For example, the phase separation process [11–15], diluents [7,8,10,16,17], solidification bath [11–14], and kinetic thermodynamics [18,19] in the TIPS have been reported on in detail. Similarly, research on co-blending grafting [20–26], stretching [27–29], annealing [30,31], and crystallization [5,9,30,32] in the MS has been conducted.

However, the limitation of the TIPS and MS [10,17,19,33] is that the PP molecule has no polar groups, so it is difficult for it to form effective secondary bonds with water molecules; therefore, the poor hydrophilicity makes the pores clog easily, resulting in a lower permeation and poor durability [11,34–38]. These effects lead to the need to perform complex post-processes on the PPHFM, such as grafting [11,35,39,40] and coating [37,38,41,42]. The dissolution-induced pores method (DIP), a neglected technology used for the preparation of PPHFMs, is not discussed in the available research papers. The DIP differs from the TIPS and MS in terms of its technical principles: it is not only simple enough to control the microstructure of the membrane, but also uncomplicated enough to combine hydrophilic modification and pore formation into one step.

In this study, the DIP method was adapted to melt-blended and extruded ethylene-vinyl alcohol copolymer (EVOH) functioning as a dissolvable part with PP; it was used to dissolve the EVOH with dimethyl sulfoxide (DMSO) to obtain hydrophilic polypropylene hollow-fiber membranes. The filtration performance of the hollow-fiber membranes was investigated with different EVOH contents, and this procedure was carried out to investigate the effect of different types of EVOH on membrane performance under optimal EVOH content conditions. The embedded factor was used to quantitatively evaluate the EVOH that was embedded in the matrix without being dissolved, and the blocked factor was used to quantitatively evaluate the EVOH that was completely blocked in the PP. Our groundbreaking research should provide a foundation for the development of the DIP method.

2. Materials and Methods

2.1. Materials and Equipment

The ethylene vinyl alcohol copolymer (EVOH, Nippon Synthetic Chemical Industry Co., Osaka, Japan) with ethylene segments of 24%, 27%, 29%, 32%, 38%, and 44% was dried under vacuum with an oven at 40 °C for 12 hours. Polypropylene was purchased from Sinopec Co., Ltd. (Beijing, China), with a melt index of 3.5 g/10 min. Dimethyl sulfoxide (DMSO, 99.9%) was used after de-watering by molecular sieve. The ink (Shanghai Hero Co., Ltd., Shanghai, China) was used to evaluate the rejection of membranes. A twin-screw extruder (TS-18) with a screw diameter of 1.8 cm and a length of 150 cm was supplied by Nanjing Huaju Co., Ltd. (Nanjing, China); the inner diameter and outer diameter of the spinneret were 1.2 mm and 1.8 mm, respectively.

2.2. Preparation of Hollow-Fiber Membranes

The EVOH (24% ethylene) and PP were mixed with a twin-screw extruder, with a ratio of EVOH from 38 wt.% to 48 wt.%. The temperature of the zones (TS-18) was 90 °C, 160 °C, 180 °C, 180 °C, and 180 °C, respectively; the host speed was 130 r/min, the winding speed was 1.2 m/s, and the stretching ratio was 6 times. The hollow fibers were cut into suitable lengths and soaked with DMSO for 24 hours; they were taken out to dry for testing and named m38 to m48 based on their content of EVOH.

The EVOH and PP were mixed with the mass of EVOH, which was 42 wt.%, wherein the ethylene content of EVOH was 24%, 27%, 29%, 32%, 38%, and 44%. The spinning-related process parameters were the same as described above, and the hollow-fiber membranes were named from M24 to M44 according to their content of ethylene.

2.3. FTIR, Particle Size/Zeta Potential, Pore Size Testing, and SEM Testing

The Fourier Transform Infrared Spectrometer (FTIR, Nicolet iS50, Madison, WI, USA) was used to obtain the molecular structure information. A nanoparticle size and zeta potential analyzer (Anton Paar, Graz, Austria) was used to evaluate the size of ink nanoparticles (Shanghai Hero Co., Ltd.). The samples of PPHFM were freeze-dried, and a scanning electron microscope (SEM SU8010 Hitachi, Minato-ku Tokyo, 1.5 kV, 10 μA) was used. The pore size distribution was tested by a liquid–liquid pore size analyzer (PSMA-10, Nanjing, China).

2.4. Embedded and Blocked Factor

The experiments were repeated 3 to 5 times, and the errors were displayed in the form of error bars.

The embedded factor is defined as the mass ratio of the substance that is not dissolved for encapsulation by matrix. The blocked factor is defined as the mass ratio of the substance that is not dissolved due to its compatibility blend with the matrix.

For example, the m40's embedded factor and blocked factor are calculated as follows:

The mass of m40 before being soaked by DMSO is m_1 (g), and the theoretical EVOH content is m_0 ($m_0 = m_1 \times 40\%$); the mass of m40 is m_2 (g). The m_3 (g) is obtained after the m40 is crushed with liquid nitrogen and then washed by DMSO and dried to obtain the mass.

The embedded factor (p) is calculated by:

$$p = \frac{m_2 - m_3}{m_0} \times 100\% \quad (1)$$

The blocked factor (b) is calculated as follows:

$$b = \frac{m_0 - (m_1 - m_3)}{m_0} \times 100\% \quad (2)$$

2.5. Filtration Performance and Porosity

The permeation of the membranes was tested with the device developed by our group (Figure 1), and calculated as follows:

$$F = \frac{V_f}{tPS} \quad (3)$$

where F (Lm^{-2}·h^{-1}·bar^{-1}) is the permeation, V_f (L) is the volume of filtrated water, P (0.1 MPa) is the pressure, S (m^2) is the area of filtration, and t (h) is the filtration time.

The rejection of ink was determined by a UV absorption spectrometer (Shimadzu 1800, λ = 307nm) via measuring absorbance with the standard curve method [43], which was calculated as follows:

$$R = \frac{A - A_0}{A} \times 100\% \quad (4)$$

where R (%) is the rejection, A is the absorbance of the unfiltered liquid, and A_0 is the absorbance of the filtered liquid.

The porosity of membranes was tested with wet method [44,45] and calculated by:

$$Pr = \frac{m_4 - m_5}{\rho_{water} V_0} \times 100\% \quad (5)$$

where Pr (%) is the porosity, V_0 (cm^3) is the membrane volume, m_4 (g) is the wet membrane mass, m_5 (g) is the dry membrane mass, and ρ_{water} (1 g/cm^3) is the water density.

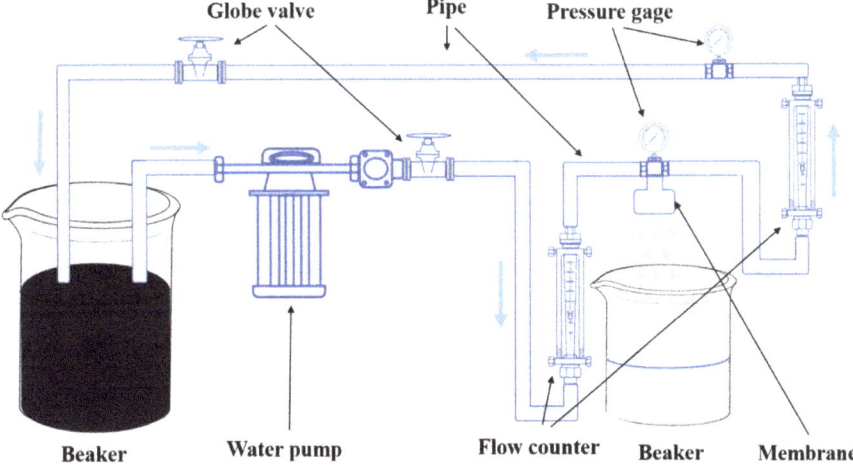

Figure 1. Schematic diagram of the hollow-fiber membrane permeation and rejection test device.

2.6. Mechanical Test

The elongation at break was tested by a tensile machine (Shenzhen Kaiqiang Co., Ltd., Shenzhen, China) and calculated by:

$$Er = \frac{\Delta L}{L_0} \times 100\% \qquad (6)$$

where Er (%) is the elongation at break, ΔL (%) is the change in sample in length at break, and L_0 (%) is the length of the sample before testing.

The strength was calculated as follows:

$$T = \frac{f}{S_0} \qquad (7)$$

where T (MPa) is the strength, f (N) is the force, and S_0 (m^2) is the cross-sectional area of the sample.

3. Results and Discussion

3.1. Preparation of Hollow-Fiber Membranes (with Different Contents of EVOH)

Figure 2 shows the surfaces of the PPHFMs prepared by the DIP method with EVOH. The PPHFMs show a unique structure of microfibers, and the diameter of the microfibers decreases as the EVOH content increases. Polymers' incompatibility is the main reason for the microfibrillation of blends after stretching [46–48], so it is possible that EVOH and PP are not fully compatible during the blending process; the shearing cuts the sizes of the PP and EVOH into small phases, and then the stretching process promotes the microfibrillation of PP. So, the increase in EVOH content promotes the progress of microfibrillation, making the microfibrils of PP more obvious, which suggests that the content of EVOH can directly control the surface of the PPHFM and greatly reduce the difficulty of controlling microfibrillation.

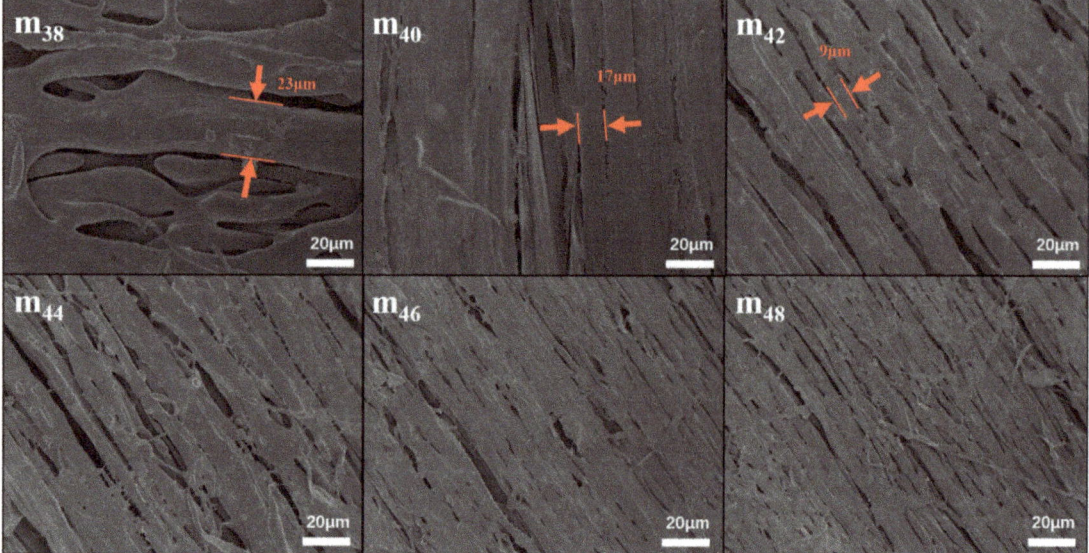

Figure 2. Surface morphology of hollow-fiber membranes with different contents of EVOH (24% ethylene).

Figure 3 shows the cross-sectional morphologies of the PPHFMs when the EVOH content is 38 wt.% and 48 wt.%, respectively. Compared with the surface topography of the PPHFMs (Figure 2), the cross-section of the m38 is similar to its surface, both of which are superimposed and compacted by microfibers with branches, but the microfibers in the surface topography show a certain orientation: the cross-section microfibers appear to be very chaotic. The m40 is also similar in terms of its surface morphology; the microfibrillation of the section is obvious. Different from the microfibers of the m38 section, the microfibers of the m40 section are more slender, and their orientation is more obvious, which is similar to the surface morphology of the m40.

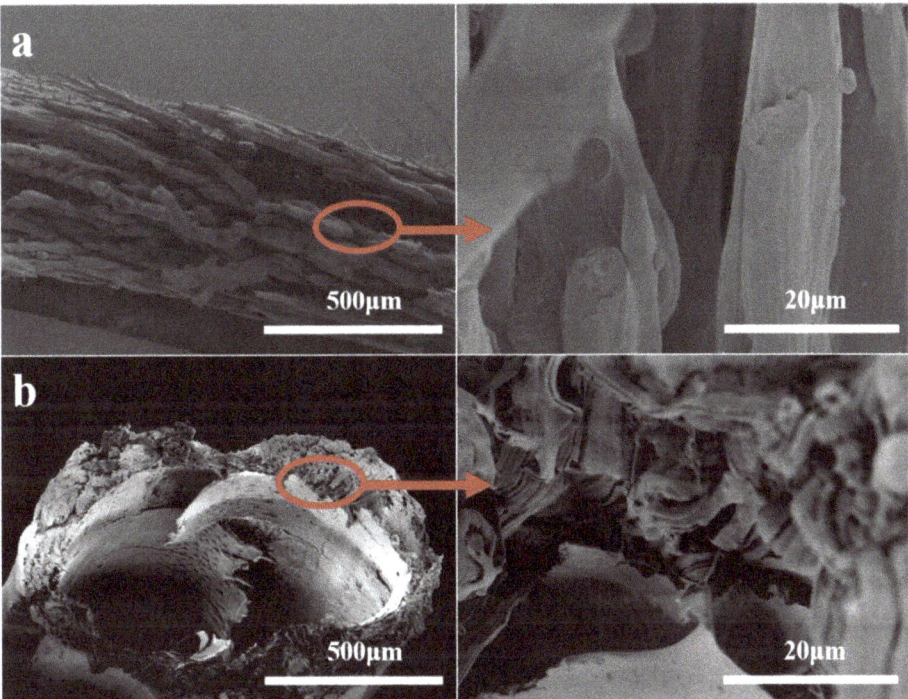

Figure 3. Cross morphology of hollow-fiber membranes m38 (**a**), m48 (**b**).

Figure 4 shows the changes in the blocked and embedded factor of EVOH in the PPHFM with different contents of EVOH. The pore-forming process of the membrane prepared by the dissolution method is also the process of the polymer being dissolved. However, it cannot be ensured that all the soluble polymers can be removed, and the soluble polymers that cannot be removed can be divided into two cases: one is that the EVOH should be dissolved, but due to the complete encapsulation of the PP, it cannot be exposed to the DMSO and cannot be removed; the other is that the soluble EVOH and the PP are continuously sheared to achieve complete miscibility. The EVOH, in this case, is also difficult to remove. In order to distinguish the above two cases, the blocked coefficient (b) and embedding coefficient (p) were defined to quantify the description. With the increase in EVOH content, the embedded factor continued to decrease, and the blocked factor continued to increase (Figure 3). When the EVOH content increased from 38 wt.% to 48 wt.%, the embedding factor decreased from 12% to 0.6%, whereas the blocked factor increased from 0.8% to 4.1%, an increase of more than five times. This shows that, in the process of increasing the EVOH content, the embedded EVOH gradually decreased, but the EVOH blocked in the PP gradually increased.

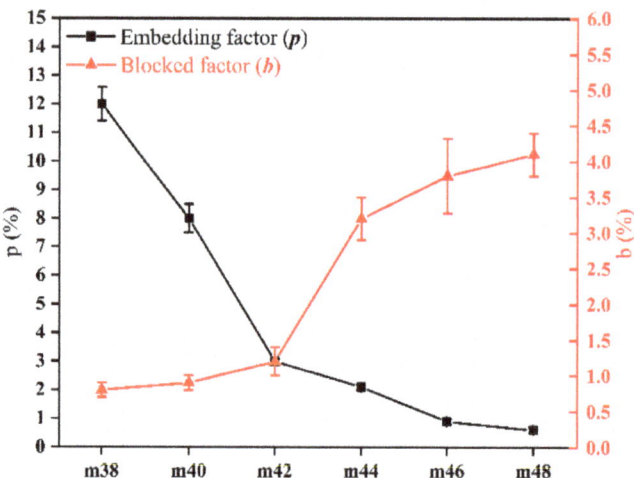

Figure 4. Blocked factor p and embedded factor b of hollow-fiber membrane with different EVOH (ethylene chain segment content of 24%) contents.

Figure 5 shows the filtration performance and porosity of the hollow-fiber membranes with different EVOH contents and the size of the particles in the ink. In general, with the increase in EVOH content, the rejection of the PPHFM showed a decreasing trend, whereas the permeation and porosity showed an increasing trend. When the EVOH content is 40 wt.%, the hollow-fiber membrane's rejection of the ink reaches 100% and the water permeation is 1300 $Lm^{-2} \cdot h^{-1} \cdot bar^{-1}$. When the EVOH content exceeds 40 wt.%, the rejection drops rapidly and the permeation rises sharply. Compared with the surface topography of the PPHFM (Figure 2), although the uniformity of the microfiber increased with the increasing EVOH content during the microfibrillation process, such topographic changes did not contribute to the rejection. The contribution of microfibrillation is an increase in permeation when the EVOH content increases, but when the EVOH content exceeds 44 wt.% the rejection drops to 0%, resulting in membrane failure. The above data also illustrate that EVOH content is the main factor controlling the membrane filtration performance. The ink is an aqueous solution of carbon particles that is used to detect the rejection performance of the membrane, and its particle size distribution is obviously concentrated (Figure 5b); the most probable particle size distribution is 141 nm.

Figure 6 shows the pore size distribution test results of the hollow-fiber membranes with an EVOH content of 38 wt.% and 40 wt.%. In the permeation and rejection tests, the ink rejection of the membranes m38 and m40 was greater than 90% (Figure 5), so these two membranes were chosen for pore size analysis for comparison and discussion. It can be seen from the results in Figure 6 that the pore size distribution range of m38 is 22–30 nm, whereas that of m40 is 80–210 nm. Obviously, compared with m40, the pore size distribution of m38 is narrower, and the membrane pore size is smaller. This shows that in the DIP method, increasing the EVOH content can expand the pores but also make the pore size distribution wider.

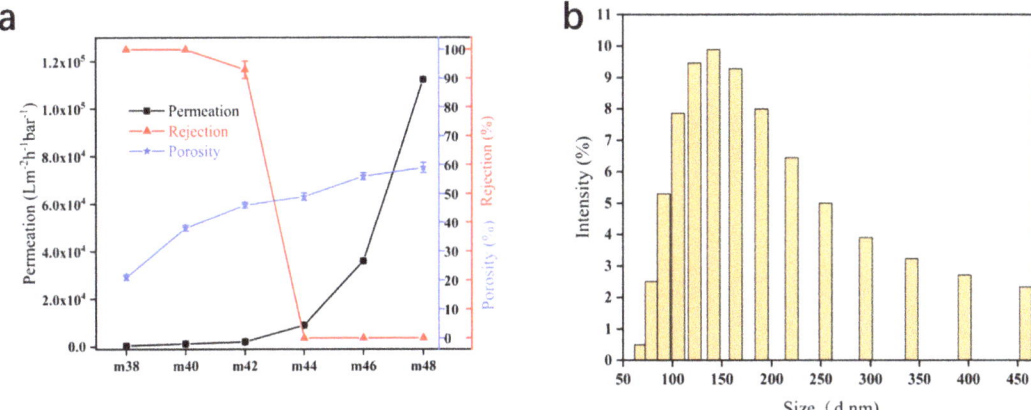

Figure 5. Changes in permeation, rejection and porosity of hollow-fiber membranes at different EVOH (ethylene chain segment content of 24%) contents (**a**) and the size and distribution of ink particles (**b**).

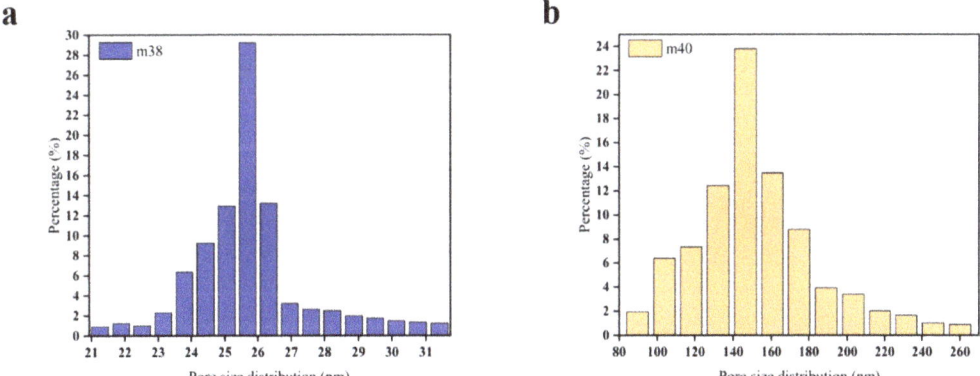

Figure 6. Pore size distribution of hollow-fiber membranes m38 (**a**) and m40 (**b**).

Figure 7 shows the mechanical properties and pure water contact angle of the hollow-fiber membranes. Good mechanical properties guarantee the membranes' service. The elongation of the membrane increases from 67% to 655% when the EVOH content increases from 38 wt.% to 48 wt.%. The strength of the PPHFM shows a different tendency; it rises from 19 MPa to 28 MPa and then falls to 24 MPa. This interesting phenomenon may come from the two changes brought about by microfibrillation: thinner fibers result in an increased elongation at break, and the thinner fibers also reduce the strength as a result of fewer microfibers sticking to each other (Figure 2). The reduction in microfiber contact points leads to an increase in the independence of microfibers during the stretching process; the lack of an overall connection and the stress caused by structural changes is transmitted to all microfibers, so the strength decreases. When the content of the EVOH is 40 wt.%, the elongation at break is 188% and the strength reached is 22 MPa. The water contact angle decreases with the increasing EVOH content, which also confirms the trend of the blocked factor b; the hydrophilicity of the PPHFM can only be inherited from the EVOH because PP is a hydrophobic material [40]. Thus, the blocked mass of the EVOH in the PP was reflected in the change in the water contact angle.

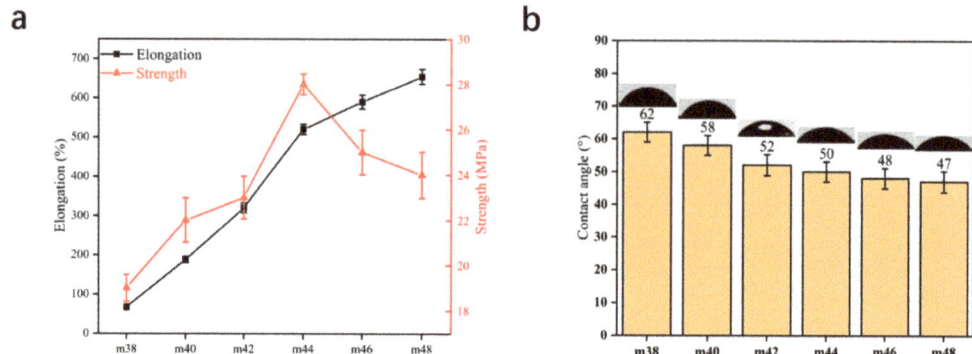

Figure 7. Mechanical properties (**a**) and contact angle (**b**) of PPHFM with different EVOH (ethylene chain segment content of 20%) contents.

3.2. Preparation of Hollow-Fiber Membranes (EVOH with Different Ethylene Segments)

The DIP method should be further investigated for the intrinsic law. Thus, the EVOH (with a mass ratio of 42 wt.%) was studied with different ethylene segment contents. Figure 8 shows that the microfibrillar structure was weakened by increasing the cross-linking of points with each other as the ethylene segment increased, and the pores were changed from extremely narrow to elliptical, which was related to the increase in the ethylene segments of the EVOH, leading to improved compatibility with PP and less dissolution of the EVOH.

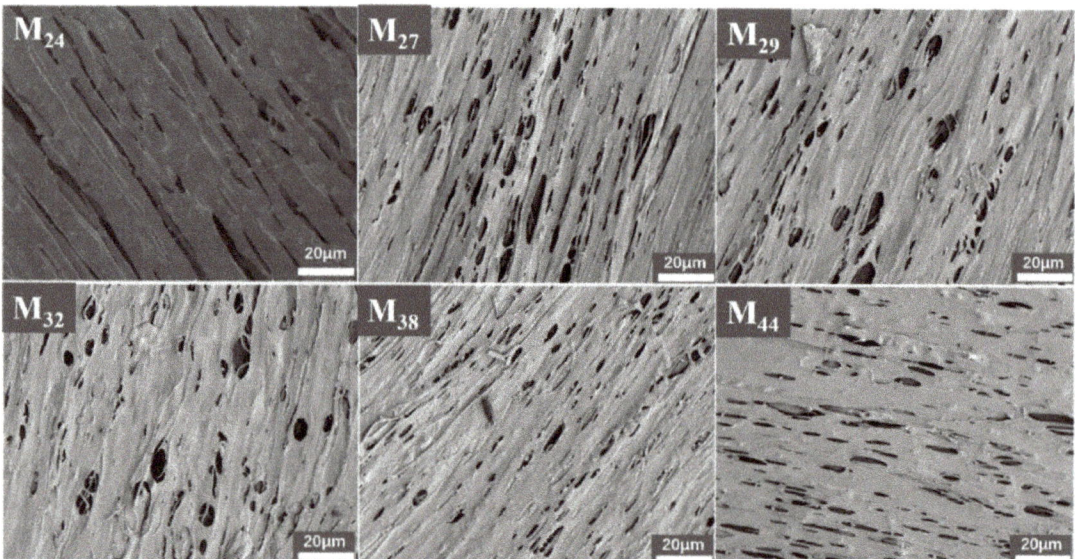

Figure 8. Surface morphology of EVOH hollow-fiber membranes with different ethylene chain segment contents.

Figure 9 shows the blocked factor and embedded factor of PPHFM with the different kinds of EVOH. It is clear that the embedding and blocking factors increased with the increase in the polyethylene chain segments of the EVOH. The embedded factor was increased from 3% to 6.8% and the blocking factor was increased from 1.2% to 11.9%. This indicates that the compatibility with the PP matrix is enhanced with the increase in the polyethylene segments of the EVOH, which is also reflected in the disappearance of microfibrils on the membrane surface (Figure 8).

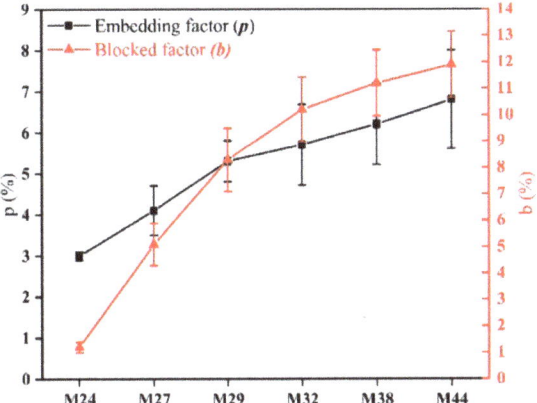

Figure 9. Blocked factor b and embedded factor p of hollow-fiber membranes (EVOH with different ethylene chain segment contents).

Figure 10a clearly illustrates that the permeation, rejection, and porosity of the membrane decreased with the increase in the EVOH's polyethylene section. The permeation of the membrane decreased from 2100 $Lm^{-2} \cdot h^{-1} \cdot bar^{-1}$ to 230 $Lm^{-2} \cdot h^{-1} \cdot bar^{-1}$, the rejection decreased from 93% to 70%, and the porosity decreased from 46% to 18%. Because the porosity decreased, the water flow was directly reduced, but at the same time as the porosity decreased, the rejection also decreased, indicating that the pore size of the membrane was increasing, as can be directly observed in Figure 8. Such changes may be due to an increase in the embedded factor (Figure 9), which reduces the porosity and increases the pore size. Thus, it is not a good choice to increase the polyethylene segment in the EVOH to improve the performance of the membrane. However, it is helpful for us to better understand the mechanism of the DIP. The increase in the blocked factor indicates that the amount of undissolved EVOH increased, which explains the decrease in porosity, and the FTIR (3305 cm^{-1}, stretching vibration of −OH) of M24 verified that some EVOH remained on the membrane (Figure 10b).

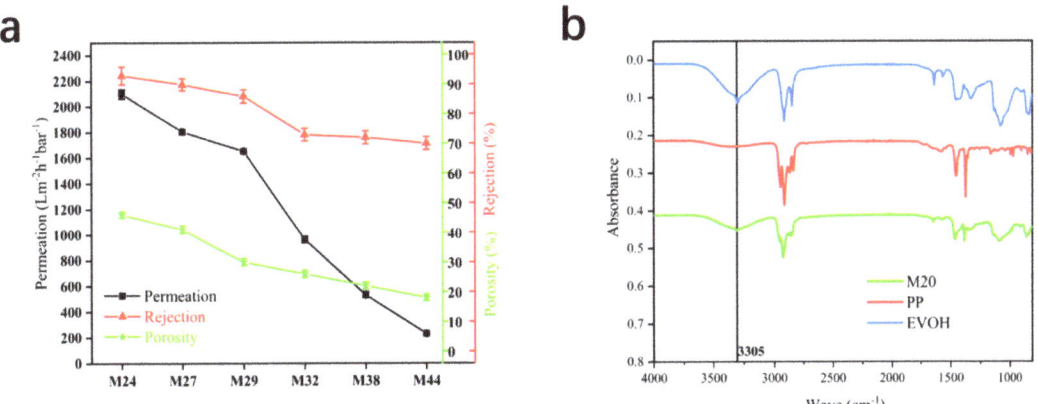

Figure 10. (**a**) Permeation, rejection, and porosity of EVOH hollow-fiber membranes with different ethylene chain segment contents, (**b**) FTIR of membrane M24, PP, and EVOH.

Figure 11 shows the mechanical properties and water contact angle of the membranes. The elongation of the membranes exhibited a gradual decreasing trend from 320% to 261%, but the strength exhibited a gradual increasing trend from 23 MPa to 28.2 MPa. Changes in mechanical properties are often related to changes in structure, and it is reasonable to conclude that the reduction in microfibers increases the strength of the membrane and reduces the elongation. Thus, it is feasible to adjust the mechanical properties of hollow-fiber membranes by controlling the polyethylene segment content of EVOH.

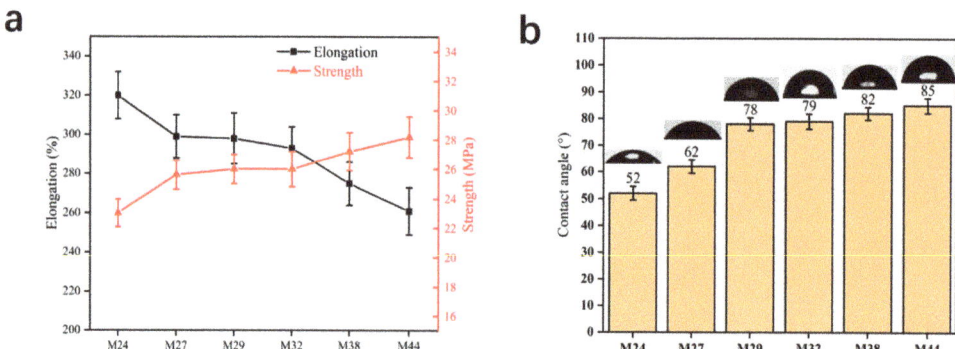

Figure 11. Mechanical properties (**a**) and water contact angle (**b**) of EVOH hollow-fiber membranes with different ethylene chain segment contents.

As concluded above, the hydrophilicity of the membrane is derived from EVOH, and the increase in the blocked factor reduces the contact angle, but the decrease in hydroxyl increases the contact angle. The results show that the contact angle has a significant increasing trend, so it is obvious that the decrease in hydroxyl plays a major role.

4. Conclusions

In this study, a hydrophilic polypropylene hollow-fiber membrane was prepared by the dissolution-induced porous method. It was found that the filtration performance of the membrane was controlled by the content and the structure of EVOH. The increase in EVOH content decreased the embedded factor, increased the blocked factor and the hydrophilicity of the membrane, made the microfibrillation more pronounced, and increased the porosity. The increase in the polyethylene segment of EVOH increased both the embedded factor and blocked factor, decreased the hydrophilicity of the membrane, caused the microfibrillation to disappear, and decreased the porosity. The preparation of PP hollow fiber prepared by the DIP method should lay the foundation for the further development of the DIP method.

Author Contributions: Conceptualization, methodology, software, validation, formal analysis, investigation, resources, data curation, writing—original draft preparation, visualization, supervision, project administration, funding acquisition Z.Q.; writing—review and editing, Z.Q. and C.H. All authors have read and agreed to the published version of the manuscript.

Funding: Supported by the Fundamental Research Funds for the Central Universities and Graduate Student Innovation Fund of Donghua University.

Institutional Review Board Statement: Not applicable.

Informed Consent Statement: Informed consent was obtained from all subjects involved in the study.

Data Availability Statement: Data is contained within the article.

Conflicts of Interest: The authors declare no conflict of interest.

References

1. Zhang, Z.; Yu, D.; Xu, X.; Li, H.; Mao, T.; Zheng, C.; Huang, J.; Yang, H.; Niu, Z.; Wu, X. A polypropylene melt-blown strategy for the facile and efficient membrane separation of oil-water mixtures. *Chin. J. Chem. Eng.* **2021**, *29*, 383–390. [CrossRef]
2. Zheng, X.; Fan, R. Covalent modification of chitosan quaternary ammonium salt on microporous polypropylene membrane and its antibacterial properties. *Chem. Ind. Eng. Prog.* **2021**, *40*, 332–338.
3. Zhu, Y.; Lu, Y.; Yu, H.; Jiang, G.; Zhao, X.; Gao, C.; Xue, L. Super-hydrophobic F-TiO$_2$@PP membranes with nano-scale'coral'-like synapses for waste oil recovery. *Sep. Purif. Technol.* **2021**, *267*, 118579. [CrossRef]
4. Yin, M.; Huang, J.; Yu, J.; Chen, G.; Qu, S.; Wang, X.; Li, C. The polypropylene membrane modified by an atmospheric pressure plasma jet as a separator for lithium-ion button battery. *Electrochim. Acta* **2018**, *260*, 489–497. [CrossRef]
5. Yang, S.; Gu, J.; Yin, Y. A biaxial stretched β-isotactic polypropylene microporous membrane for lithium-ion batteries. *J. Appl. Polym. Sci.* **2018**, *135*, 45825. [CrossRef]
6. Cornelissen, C.G.; Dietrich, M.; Gromann, K.; Frese, J.; Krueger, S.; Sachweh, J.S.; Jockenhoevel, S. Fibronectin coating of oxygenator membranes enhances endothelial cell attachment. *Biomed. Eng. Online* **2013**, *12*, 7. [CrossRef]
7. Takahashi, A.; Tatebe, K.; Onishi, M.; Seita, Y.; Takahara, K. Influence of Molecular-Weight of Polypropylene and A Nuclenting-Agent on Polypropylene Miroporous Hollow-Fiber Membranes for Artificial Lungs. *Kobunshi Ronbunshu* **1993**, *50*, 507–513. [CrossRef]
8. Takahashi, A.; Tatebe, K.; Onishi, M.; Seita, Y.; Takahara, K. Morphological Change of Microporous Hollow-Fiber Membranes for Artificial Lungs Induced by Cooling. *Kobunshi Ronbunshu* **1993**, *50*, 515–521. [CrossRef]
9. Shao, H.; Wei, F.; Wu, B.; Zhang, K.; Yao, Y.; Liang, S.; Qin, S. Effects of annealing stress field on the structure and properties of polypropylene hollow fiber membranes made by stretching. *RSC Adv.* **2016**, *6*, 4271–4279. [CrossRef]
10. Yan, S.-Y.; Wang, Y.-J.; Mao, H.; Zhao, Z.-P. Fabrication of PP hollow fiber membrane via TIPS using environmentally friendly diluents and its CO$_2$ degassing performance. *RSC Adv.* **2019**, *9*, 19164–19170. [CrossRef]
11. Wang, S.; Zhang, X.; Xi, Z.; Wang, Y.; Qiao, J. Design and preparation of polypropylene ultrafiltration membrane with ultrahigh permeation for both water and oil. *Sep. Purif. Technol.* **2020**, *238*, 116455. [CrossRef]
12. Amirabedi, P.; Akbari, A.; Yegani, R. Fabrication of hydrophobic PP/CH$_3$SiO$_2$ composite hollow fiber membrane for membrane contactor application. *Sep. Purif. Technol.* **2019**, *228*, 115689. [CrossRef]
13. Milad, F.; Habib, E.; Reza, Y.; Saber, Z. Fouling characterization of TiO$_2$ nanoparticle embedded polypropylene membrane in oil refinery wastewater treatment using membrane bioreactor (MBR). *Desalination Water Treat.* **2017**, *90*, 99–109. [CrossRef]
14. Taghaddosi, S.; Akbari, A.; Yegani, R. Preparation, characterization and anti-fouling properties of nanoclays embedded polypropylene mixed matrix membranes. *Chem. Eng. Res. Des.* **2017**, *125*, 35–45. [CrossRef]
15. Tang, N.; Li, Z.; Hua, X. Study on structure and hydrophobicity of PP/EVA co-blending membrane: Quenching rate. In *AIP Conference Proceedings*; AIP Publishing LLC: Melville, NY, USA, 2017.
16. Zwirner, U.; Hoeffler, K.; Pflaum, M.; Korossis, S.; Haverich, A.; Wiegmann, B. Identifying an optimal seeding protocol and endothelial cell substrate for biohybrid lung development. *J. Tissue Eng. Regen. Med.* **2018**, *12*, 2319–2330. [CrossRef]
17. Wang, Y.-J.; Zhao, Z.-P.; Xi, Z.-Y.; Yan, S.-Y. Microporous polypropylene membrane prepared via TIPS using environment-friendly binary diluents and its VMD performance. *J. Membr. Sci.* **2018**, *548*, 332–344. [CrossRef]
18. Wang, Y.J.; Yan, S.Y.; Zhao, Z.P.; Xi, Z.Y. Isothermal Crystallization of iPP in Environment-friendly Diluents: Effect of Binary Diluents and Crystallization Temperature on Crystallization Kinetics. *Chin. J. Polym. Sci.* **2019**, *37*, 617–626. [CrossRef]
19. Hao, J.; Fan, Z.; Xiao, C.; Zhao, J.; Liu, H.; Chen, L. Effect of stretching on continuous oil/water separation performance of polypropylene hollow fiber membrane. *Iran. Polym. J.* **2017**, *26*, 941–948. [CrossRef]
20. Feng, J.; Zhang, G.; MacInnis, K.; Olah, A.; Baer, E. Structure-property relationships of microporous membranes produced by biaxial orientation of compatibilized PP/Nylon 6 blends. *Polymer* **2018**, *145*, 148–156. [CrossRef]
21. Jiang, S.; Wang, W.; Ding, Y.; Yu, Q.; Yao, L. Preparation and characterization of antibacterial microporous membranes fabricated by poly(AMS-co-DMAEMA) grafted polypropylene via melt-stretching method. *Chin. Chem. Lett.* **2018**, *29*, 390–394. [CrossRef]
22. Jiang, S.; Wang, W.; Huang, T.; Ma, J.; Ding, Y.; Yu, Q. Microporous membrane fabricated by AMS-GMA-TPE terpolymer grafted polypropylene prepared via extrusion. *J. Appl. Polym. Sci.* **2018**, *135*, 46020–46025. [CrossRef]
23. Wang, S.; Ajji, A.; Guo, S.; Xiong, C. Preparation of Microporous Polypropylene/Titanium Dioxide Composite Membranes with Enhanced Electrolyte Uptake Capability via Melt Extruding and Stretching. *Polymers* **2017**, *9*, 110. [CrossRef] [PubMed]
24. Li, T.; Dai, Y.; Li, J.; Guo, S.; Xie, G. A high-barrier PP/EVOH membrane prepared through the multistage biaxial-stretching extrusion. *J. Appl. Polym. Sci.* **2017**, *134*, 45016–45027. [CrossRef]
25. Feng, J.; Zhang, G.; MacInnis, K.; Olah, A.; Baer, E. Formation of microporous membranes by biaxial orientation of compatibilized PP/Nylon 6 blends. *Polymer* **2017**, *123*, 301–310. [CrossRef]
26. Chen, H.; Ma, W.; Xia, Y.; Gu, Y.; Cao, Z.; Liu, C.; Yang, H.; Tao, S.; Geng, H.; Tao, G.; et al. Improving amphiphilic polypropylenes by grafting poly(vinylpyrrolidone) and poly(ethylene glycol) methacrylate segments on a polypropylene microporous membrane. *Appl. Surf. Sci.* **2017**, *419*, 259–268. [CrossRef]
27. Liu, X.D.; Ni, L.; Zhang, Y.-F.; Liu, Z.; Feng, X.-S.; Ji, L. Technology study of polypropylene hollow fiber membranes-like artificial lung made by the melt-spinning and cold-stretching method. In *Materials Processing Technology*; Advanced Materials Research; Liu, X.H., Jiang, Z., Han, J.T., Eds.; Elsevier: Amsterdam, The Netherlands, 2012; Volume 418–420, pp. 26–29.

28. Liu, Z.Y.; Wu, X.T.; Yan, J.; Yang, W.; Yang, M.B. Effect of Annealing Temperature on PP Microporous Membranes Obtained by a Melt-Extrusion-Stretching Method. *Int. Polym. Proc.* **2019**, *34*, 467–474. [CrossRef]
29. Luo, D.; Wei, F.; Shao, H.; Xie, L.; Cui, Z.; Qin, S.; Yu, J. Microstructure construction of polypropylene-based hollow fiber membranes with bimodal microporous structure for water permeation enhancement and rejection performance rejection. *Sep. Purif. Technol.* **2019**, *213*, 328–338. [CrossRef]
30. Han, S.W.; Woo, S.M.; Kim, D.J.; Park, O.O.; Nam, S.Y. Effect of annealing on the morphology of porous polypropylene hollow fiber membranes. *Macromol. Res.* **2014**, *22*, 618–623. [CrossRef]
31. Mei, L.; Zhang, D.; Wang, Q. Morphology structure study of polypropylene hollow fiber membrane made by the blend-spinning and cold-stretching method. *J. Appl. Polym. Sci.* **2002**, *84*, 1390–1394. [CrossRef]
32. Ruijie, X.; Jiayi, X.; Ziqin, T.; Henghui, H.; Xiande, C.; Caihong, L.; Xingqi, Z. Pore growth and stabilization in uniaxial stretching polypropylene microporous membrane processed by heat-setting. *J. Polym. Sci. Part B Polym. Phys.* **2018**, *56*, 1604–1614. [CrossRef]
33. Kim, B.T.; Song, K.; Kim, S.S. Effects of Nucleating Agents on Preparation of Polypropylene Hollow Fiber Membranes by Melt Spinning Proces. *Macromol. Res.* **2002**, *10*, 127–134. [CrossRef]
34. Jin, J.; Zhang, K.; Du, X.; Yang, J. Synthesis of polydopamine-mediated PP hollow fibrous membranes with good hydrophilicity and antifouling properties. *J. Appl. Polym. Sci.* **2017**, *134*. [CrossRef]
35. Shao, H.; Qi, Y.; Luo, D.; Liang, S.; Qin, S.; Yu, J. Fabrication of antifouling polypropylene hollow fiber membrane breaking through the selectivity-permeability trade-off. *Eur. Polym. J.* **2018**, *105*, 469–477. [CrossRef]
36. Le, T.-N.; Au-Duong, A.-N.; Lee, C.-K. Facile coating on microporous polypropylene membrane for antifouling microfiltration using comb-shaped poly(N-vinylpyrrolidone) with multivalent catechol. *J. Membr. Sci.* **2019**, *574*, 164–173. [CrossRef]
37. Wardani, A.K.; Ariono, D.; Subagjo, S.; Wenten, I.G. Fouling tendency of PDA/PVP surface modified PP membrane. *Surf. Interfaces* **2020**, *19*, 100464. [CrossRef]
38. Wenten, I.G.; Khoiruddin, K.; Wardani, A.K.; Aryanti, P.T.P.; Astuti, D.I.; Komaladewi, A.A.I.A.S. Preparation of antifouling polypropylene/ZnO composite hollow fiber membrane by dip-coating method for peat water treatment. *J. Water Process Eng.* **2020**, *34*, 101158. [CrossRef]
39. Shao, H.; Qi, Y.; Liang, S.; Qin, S.; Yu, J. Interface engineering of polypropylene hollow fiber membrane through ultrasonic capillary effect and nucleophilic substitution. *Polym. Adv. Technol.* **2018**, *29*, 3125–3133. [CrossRef]
40. Shao, H.; Qi, Y.; Liang, S.; Qin, S.; Yu, J. Polypropylene composite hollow fiber ultrafiltration membranes with an acrylic hydrogel surface by in situ ultrasonic wave-assisted polymerization for dye removal. *J. Appl. Polym. Sci.* **2018**, *136*, 47099–47109. [CrossRef]
41. Li, N.; Chen, G.; Zhao, J.; Yan, B.; Cheng, Z.; Meng, L.; Chen, V. Self-cleaning PDA/ZIF-67@PP membrane for dye wastewater remediation with peroxymonosulfate and visible light activation. *J. Membr. Sci.* **2019**, *591*, 117341. [CrossRef]
42. Liu, M.; Wu, Y.; Wu, Y.; Gao, M.; Lü, Z.; Yu, S.; Gao, C. Cross-flow deposited hydroxyethyl cellulose (HEC)/polypropylene (PP) thin-film composite membrane for aqueous and non-aqueous nanofiltration. *Chem. Eng. Res. Des.* **2020**, *153*, 572–581. [CrossRef]
43. Jankowski, J.A.; Ernst, T.; Sucksdorff, C.; Pirjola, R.; Ryno, J. Experiences of a filter method and a standard curve method for determining k-indices. *Ann. Geophys. Atmos. Hydrospheres Space Sci.* **1988**, *6*, 589–593.
44. Gryta, M. Influence of polypropylene membrane surface porosity on the performance of membrane distillation process. *J. Membr. Sci.* **2007**, *287*, 67–78. [CrossRef]
45. Zhang, Z.; Gao, J.; Zhang, W.; Ren, Z. Experimental study of the effect of membrane porosity on membrane absorption process. *Sep. Sci. Technol.* **2006**, *41*, 3245–3263. [CrossRef]
46. Yang, R.; Chen, L.; Zhang, W.-Q.; Chen, H.-B.; Wang, Y.-Z. In situ reinforced and flame-retarded polycarbonate by a novel phosphorus-containing thermotropic liquid crystalline copolyester. *Polymer* **2011**, *52*, 4150–4157. [CrossRef]
47. Shahnooshi, M.; Javadi, A.; Nazockdast, H.; Ottermann, K.; Altstädt, V. Rheological rationalization of in situ nanofibrillar structure development: Tailoring of nanohybrid shish-kebab superstructures of poly (lactic acid) crystalline phase. *Polymer* **2020**, *211*, 123040. [CrossRef]
48. Sakai, Y.; Umetsu, K.; Miyasaka, K. Mechanical properties of biaxially drawn films of ultra-high molecular weight polyethylene dried gels. *Polymer* **1993**, *34*, 318–322. [CrossRef]

MDPI
St. Alban-Anlage 66
4052 Basel
Switzerland
www.mdpi.com

MDPI Books Editorial Office
E-mail: books@mdpi.com
www.mdpi.com/books

Disclaimer/Publisher's Note: The statements, opinions and data contained in all publications are solely those of the individual author(s) and contributor(s) and not of MDPI and/or the editor(s). MDPI and/or the editor(s) disclaim responsibility for any injury to people or property resulting from any ideas, methods, instructions or products referred to in the content.

www.ingramcontent.com/pod-product-compliance
Lightning Source LLC
LaVergne TN
LVHW070623100526
838202LV00012B/712